A Compendium of
British Mining

*With Statistical Notices of
the Principal Mines in Cornwall*

Joseph Yelloly Watson

CAMBRIDGE
UNIVERSITY PRESS

CAMBRIDGE
UNIVERSITY PRESS

University Printing House, Cambridge, CB2 8BS, United Kingdom

Published in the United States of America by Cambridge University Press, New York

Cambridge University Press is part of the University of Cambridge.
It furthers the University's mission by disseminating knowledge in the pursuit of
education, learning and research at the highest international levels of excellence.

www.cambridge.org
Information on this title: www.cambridge.org/9781108063692

© in this compilation Cambridge University Press 2013

This edition first published 1843
This digitally printed version 2013

ISBN 978-1-108-06369-2 Paperback

CAMBRIDGE LIBRARY COLLECTION

Books of enduring scholarly value

Technology

The focus of this series is engineering, broadly construed. It covers technological innovation from a range of periods and cultures, but centres on the technological achievements of the industrial era in the West, particularly in the nineteenth century, as understood by their contemporaries. Infra-structure is one major focus, covering the building of railways and canals, bridges and tunnels, land drainage, the laying of submarine cables, and the construction of docks and lighthouses. Other key topics include develop-ments in industrial and manufacturing fields such as mining technology, the production of iron and steel, the use of steam power, and chemical processes such as photography and textile dyes.

A Compendium of British Mining

Joseph Yelloly Watson (1817–88) produced this short work for private circu-lation in 1843. For many years a mining agent with the London firm of Watson and Cuell, he became a fellow of the Geological Society and wrote on mining and historical subjects. Drawing chiefly on data from Cornwall, the present work gives details of mining processes, the layout of mines, the working conditions of miners (including figures for wages and working hours), and the typical management structure of a mine, with information on shareholders, profits and accounting. The work includes coverage of individual mining districts, including Gwennap, Camborne, Illogan, St Just, St Ives and Marazion in Cornwall, along with discussion of mines in Devon, Cumberland and elsewhere. Concluding with notes on the history of metalmining, followed by a useful glossary of mining terms, this remains a concise and instructive resource on a significant economic sector in the nineteenth century.

Cambridge University Press has long been a pioneer in the reissuing of out-of-print titles from its own backlist, producing digital reprints of books that are still sought after by scholars and students but could not be reprinted economically using traditional technology. The Cambridge Library Collection extends this activity to a wider range of books which are still of importance to researchers and professionals, either for the source material they contain, or as landmarks in the history of their academic discipline.

Drawing from the world-renowned collections in the Cambridge University Library and other partner libraries, and guided by the advice of experts in each subject area, Cambridge University Press is using state-of-the-art scanning machines in its own Printing House to capture the content of each book selected for inclusion. The files are processed to give a consistently clear, crisp image, and the books finished to the high quality standard for which the Press is recognised around the world. The latest print-on-demand technology ensures that the books will remain available indefinitely, and that orders for single or multiple copies can quickly be supplied.

The Cambridge Library Collection brings back to life books of enduring scholarly value (including out-of-copyright works originally issued by other publishers) across a wide range of disciplines in the humanities and social sciences and in science and technology.

A

COMPENDIUM

OF

BRITISH MINING,

WITH

STATISTICAL NOTICES

OF THE

PRINCIPAL MINES IN CORNWALL;

TO WHICH IS ADDED,

THE HISTORY AND USES OF METALS,

AND

A GLOSSARY

OF THE

TERMS AND USAGES OF MINING.

———

Compiled for the use of persons interested but not conversant with the subject.

———

By JOSEPH YELLOLY WATSON.

———

London:

PRINTED FOR PRIVATE CIRCULATION.

1843.

TABLE OF CONTENTS.

THE MINES OF MARAZION.

THE EASTERN DISTRICT OF CORNWALL AND DEVON.

THE MINES OF CUMBERLAND AND OTHER COUNTIES.

THE HISTORY OF METALS AND THEIR USES.

INTRODUCTION.

'Tis said, that " a knowledge of our subterranean wealth would be the means of furnishing greater opulence to the country, than the acquisition of the mines of Mexico and Péru."

THE subject of British Mining ought to be deeply interesting to the capitalist, as some millions are annually employed in this pursuit.

A prejudice which has prevailed, as to its being extremely hazardous, (perhaps arising from the enormous sums that have been wasted), is principally the result of rashness, the absence of proper information, and the want of proper attention to the pecuniary conduct of the concerns, especially in large proprietaries, where the losses have principally fallen.

But the want of information is difficult to be supplied, as there is no subject so little understood in practice and detail, by the generality of persons interested, as mining.

The press, which furnishes so profusely, matter, both elementary and practical, on almost every subject, is so barren on this head, that even a manual of humble pretensions, is still a desideratum, and which this compilation is intended to supply.

Previous to the valuable essay on Cornish mines by Mr. Abbott in 1833, there was nothing extant but the old standard work of Pryce, so costly, as well as eagerly sought after, and which, however well describing the mining operations of " olden time," is inapplicable to the present day.

It was, perhaps, owing to this circumstance (the want of information) that capital for mining purposes, was forty or fifty years ago, furnished from only a few mercantile towns and seaports, besides Cornwall itself, while the great metropolis was almost excluded from all participation. There might be also some other reasons, for the Cornishmen of those days, got a bad name, and were deemed as " cannie as Yorkshire !" But those times are past, and it may be concluded that Cornishmen and out-adventurers now understand their own interests too well, not to harmonize.

B

But even Mr. Abbott's work, which was privately circulated, important as it was, made no allusion to the practical details, gave no explanation of the mining technicals, and was, therefore, only useful to those who were already well acquainted with the subject. Besides which, since his work was published, many mines have been abandoned, and others not then known, have come into full operation.

While alluding to the press, however, the services of the *Mining Journal*, (the earliest devoted to the subject), must not be forgotten, which has proved, perhaps, as valuable and useful to the mining interest, as the *Lancet* did to the medical profession; acting either as a scarifier when too great a plethora of mining bubbles prevailed, or as a styptic, where dangerous "bleeding," of the pockets of the capitalist was concerned. There are other periodicals, also, particularly the *Railway Magazine and Commercial Journal*, which take regular notice of mining affairs, and watch over the mining interest. In giving the following sketch of British mining, it is intended to be confined principally to that of Cornwall, as from the richness and number of its mines, and the skill employed in working them, the attention of capitalists has been turned to that county, to the comparative neglect, however, of most others.

By those acquainted with legitimate Cornish mines, it will be allowed, that when properly conducted, mining is neither so uncertain, or speculative as many have been led, either through ignorance or prejudice, to imagine; on the contrary, it frequently offers the most profitable advantages to the capitalist, great or small.

To enter indiscriminately into mining speculations without regard to locality or choice, in the faint hope that one or two may turn up prizes, and redeem those that fail, is certainly a dangerous adventure; for a judicious selection is as necessary as the skill and judgment of an effective management; but, as *all* mines, however well selected, cannot be expected to prove profitable, it is, therefore, advisable to invest in more than one undertaking; a division of risk, necessarily diminishing the effects of a partial loss, and insuring a greater degree of success on the aggregate of undertakings.

Great coolness and patience are also required in entering upon these speculations; for many have, from mere temporary causes of depression, arising from no fault of the mine, become frightened, and sold out at a heavy loss, when by remaining quiet they might have eventually realised large profits. There may be depressions also in the market for ore, when it is advisable not to sell the usual quantity per month (though it may lessen the usual dividend) until a reaction takes place; but this should not, as it too often is, be considered as a falling off in the mine, for in the well-conducted, the ticketings must not be solely looked to as the criterion of the state of the mine.

In order to illustrate the foregoing remarks, instances of fortunate and judicious selection, as well as cases in which patience and perseverance, founded on sound indications, leading to the happiest results, will be noticed. The East Pool mine may be considered as an instance of judicious selection, whether as regards the locality or the surrounding district. This mine upon an original outlay by the adventurers of only £640 has returned upwards of £100,000! and is still making large profits. The Tresavean mine is another instance of perseverance, for after having been even once or twice abandoned, it was at last resumed by a few skilful fortunate adventurers, and has made them a clear profit of £800,000! Many others might be mentioned equally prosperous; but as they will come under the head of mining notices, the above notice will suffice for the purpose here.

There are about 112 copper mines working in Cornwall, employing nearly 60,000 persons; the tin mines employ about 12,000 more. The amount annually expended in labour is estimated at £900,000; for materials for working the mines, £300,000; and the mining districts, either directly or indirectly, give employment to upwards of 100,000 individuals. The returns of Cornish copper mines for fourteen years ending 1841, were £13,682,810, the last seven years of that time averaging £1,049,821 annually. The tin mines return about £300,000 a-year.

Previous to the year 1700, it is supposed that the copper ore produced in Cornwall, was principally from the tin mines, or mines originally wrought for tin, and it was not till a much later period that mines were set to work purposely for raising copper. Pryce informs us that from 1726 to 1735, the average annual produce of copper ore exceeded 6,000 tons; in 1770, the quantity of ore had increased to 27,000 tons, which yielded according to existing documents, about 3,200 tons of copper. In 1798, the total value of copper sold in Cornwall, for that year was £405,488. 15s. 6d.; the labour amounted to £253,601. 12s. 3d.; materials, £146,253. 16s. 3d.; total cost,

£408,240. 7s. 4d. ; shewing a *loss* on that year of £2,759. 12s. 2d., the average standard of the year was £103. 12s. 6d., and the price of manufactured copper 1s. 2d. per lb. The increase of the annual returns to the present·time may be seen as above. In 1768, the immense riches of the Anglesea mines were discovered. These mines, about 1784, produced 3,000 tons of copper annually. In 1780, the singular mass of copper óre at Ecton Hill was discovered ; but these mines have ceased to be so productive, whilst those of Cornwall have greatly increased in value.

It may be noticed here, and by those acquainted with mining, the remark will be verified, that there is an extreme difficulty in obtaining statistical information from the different mines. Unfortunately for themselves, Cornishmen seem to think they thrive best in mystery ; and thus, one of the finest fields for speculation, is often looked upon by capitalists, with prejudice and distrust, when, if every transaction were fairly elucidated, and openness and candour practised, they would be met in a kindred spirit, and mining would become far more popular than it is ; and who would be benefitted more than themselves ? In collecting materials for this book, considerable difficulties in getting accurate accounts of the mines have been met with ; but through the kindness of friends, fuller particulars have been obtained than were ever collected before. And, if any of them should be inaccurate, it is the fault of those, who cast a veil over their proceedings, and view with jealousy and distrust any attempt to obtain an insight into their doings.

In compiling this work, the object has not been to enter into scientific theory or speculation, nor to assist the *practical* miner, but to give plain facts as they exist, and to endeavour to throw light upon a subject in which so many are interested, but which so few understand, and that, in as brief and lucid a manner as possible.

GENERAL FEATURES OF A MINE.

A MINE is a depository of mineral, or ore, in the bowels of the earth, and opened for the purpose of obtaining the produce. It has been legally determined that no mine can properly be said to exist, before it has been opened by shafts, pits, or levels ; for before that has been done, there can be no positive certainty that any mineral, or ore, lies in that particular district. As land includes in general everything beneath its surface, the owner in fee of land is almost invariably the owner of the mines lying underneath, with the exception of gold and silver mines, which belong by prerogative to the crown.

The mines in Cornwall are generally worked by a company of pro- prietors, called adventurers, who agree with the *lord*, or owner of the mine, for a certain number of years, paying either a fixed per centage, or a certain proportion of the ores raised, called *dues*, being 1-15th, 1-18th, or 1-20th, as may be agreed upon. The grant thus made is called a *sett*. The bounds or limits of a mine are marked on the surface by large stones, placed at equal distances. The property of the soil above being entirely distinct, but the lessees of the sett have the privi- lege or right of sinking such shafts as may be necessary for the effectual working of the mine. In commencing from the surface, a vein, or por- tion of a lode containing ore, is seldom met with at a less depth than 10 or 20 fathoms from the surface. A perpendicular shaft is first sunk, and at about 10 fathoms in depth, a horizontal level or gallery is driven by two sets of men, working in opposite directions, the ores and stuff being raised by a windlass. When this level is driven about 50 fathoms, two shafts are made at either extremity for airing the mine ; this level can be carried on to any extent. The engine shaft being sunk deeper, similar levels are driven at every 10 fathoms in depth, the shaft being always sunk to a greater depth than the lowest level. The mine being thus divided into right angled masses of 50 fathoms in length, and 10 in height, these masses are again sub-divided by small perpendicular shafts or winzes, of about 10 fathoms in height, and 16 in length ; the mine being thus finally divided into *pitches*. In addition to these shafts, a level, called the adit, is formed to drain the water from the lowest part of the mine ; the adit will carry off the water without the aid of machinery, so long as the lowest shaft is above the level of the sea, but otherwise a steam engine is required to pump up the water to a level with the adit. The great Cornish adit, which commences in a valley above Carnon, receives the branch adits of 50 mines; in the parish of Gwennap, forming excavations 30 miles in length, and in some places 400 feet below the surface of the ground. The longest branch is from Cardrew mine, being $5\frac{1}{2}$ miles in length. The great adit opens into the sea at Restronget Creek, and empties them all into Falmouth Harbour.

The valleys of Cornwall are not deep, and there are few instances where one is raised from or above the adit level, to pay for the future operations of the mines, and seldom more than a 40 fathom adit can be obtained.

Shafts are generally timbered to about 20 or 30 feet, and sometimes all the way. Levels are about 3 feet wide, and 6 or 7 feet high, and cut in the body of the vein. After the shaft has passed through the lode, and the first level run, as the lode descends with an inclined plane or dip, more or less, in order to find it at the next descent, the shaft is continued, and cross cuts made every ten fathoms to reach the lode, which being divided into *pitches*, each pitch is let to a tributer, who with his *para*, or gang, break, raise, and pay for dressing the ores ; the weekly or monthly produce being made into heaps of about 100 tons each. Samples of it are sent to assayers to determine the value according to the produce, or quantity, of fine copper contained in 100 parts of ore, and the samplings are then sold at the weekly ticketings, and the tributers receive a certain share of the value of the ores for their labour.

The estimated rate of wages for the county is as follows.—Per month :—Tributers £2 15s. to £3. 11s. 7d. ; tutworkmen £2 10s. to £3 1s. 11d. ; surface labourers £2 2s. to £2 5s. ; boys 13s. to £1 8s. ; and females 12s. to 18s. And the proportions in a hundred persons employed in a mine, is 30 tributers, 20 tutworkmen, 10 surface labourers, 25 boys, and 15 labourers. The tributer may not for many months, earn a remunerating profit, but if the indications of the lode be favourable, he will at every setting renew his bargain, in the hope that the lode may eventually become rich. If before the completion of his term, his expectations be realised, he and his *para*, or gang, are often able to work out ore to the value of £60, or £100 each, sometimes more ; but at the next renewal, the rate of tribute is re-adjusted, and fair wages earned until the ore fails.

Most rocks are traversed by fissures, and which when they contain minerals, are called *veins*, *lodes*, or *courses*. The metal contained in these veins is generally found combined with other substances, and is, therefore, called *ore*. Veins or lodes run to a considerable extent, sometimes for several miles, and have in no instance been followed to an actual termination, being always relinquished when no longer worth working ; their direction downwards generally forms an angle of 70 deg. or 80 deg. If a lode continues straight, it is called a regular lode ; (1) if it occasionally swells and contracts, an irregular lode, or a pipe vein ; (2) the wider parts are called *bunches* ; (3) and when it divides into branches it is said to *take horse*, or come into dead ground leaving a branch of ore on either side. (4.) When a vein *takes horse,* it is generally considered a good indication, for (as the miners say) at the tail of the horse, there are generally some rich bunches of ore. Sometimes a vein called a *cross course* interferes, and *heaves* the regular lode, from 2 feet to 50 fathoms, out of its course ; (5) or it becomes reduced to a mere thread, and reappears at a distance.

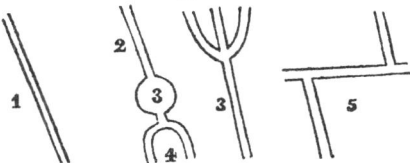

A cross lode in Wheal Peever, about three miles east of Redruth, extends from sea to sea. On its west side every vein it passes is heaved 50 fathoms farther north, from the line it would have otherwise pursued, and which the other part still keeps. It was not until after a search, during 40 years, that this heaved lode was discovered.

The most abundant substance in veins is crystallized spar, termed veinstone, or the leader of the lode ; the veins are distinguished by names, according to the nature of the veinstones. The following are the principal :—1. *Gossany*, when the veinstone is clay, mixed with silica, and oxide of iron. Its colour varies from light yellow to deep brown. This is the most common veinstone, and is considered as promising both for copper and tin. 2. *Sparry*, when quartz predominates. It is rather unpromising. 3. *Mundicky*, when iron pyrites abounds. It is considered as rather promising. 4. *Peachy*, when the veinstone is chlorite. It is more promising for tin than copper. 5. *Flookany*, when one or both of its sides is lined with bluish white clay. 6. *Caply*, when the veinstone is a hard substance of a greenish or brownish colour, chiefly a mixture of chlorite and quartz. Tin is found in it; but seldom copper. 7. *Pryany*, when the one is found in detached lumps. 8. When a vein abounds in blende it is called a *black jack lode* ; when it contains granite it is called a *growan lode*. Tin and copper lodes generally run east and west, and lead lodes north and south.

The veins in Cornwall have no determinate size, being sometimes very narrow, or exceeding several fathoms in width ; extending sometimes to a great length and depth, or terminating after a short course in either direction. As regards their form, they are occasionally, though rarely, contained within parallel and regularly inclined sides or walls ; but are continually varying in width, both on the line of their course and of their inclination, partaking often of the same undulating, and even curved form of the rocks which they traverse : moreover, they are accompanied on either side by innumerable branches, which extend in various directions. And, lastly; a parallel series of veins frequently meets a cross vein, either on the line of its course, or of its dip ; some of these veins continue their direction on either side of the cross-vein, whilst others, on the opposite side of the cross-vein, abruptly disappear on the line of their original course, and are often found at some distance therefrom, but running in a parallel direction.

Veins vary very much in their composition : in general they consist entirely of earthy minerals, which, indeed, even when the veins are metalliferous, constitute the greater part thereof, the ores seldom being continuous for any considerable distance, but being scattered and disseminated throughout the matrix in short irregular veins, layers, branches, granules, crystals, and smaller forms ; sometimes, indeed, but rarely, except in very small veins, the ore entirely prevails.

On the kindly appearance of lodes, Mr. Henwood says, " All the harder rocks in the mining districts are quartzose, and whether they are granite, elvan, or slate, this character is unfavourable. A distinctly crystalline structure of granite, and their slaty texture, and high inclination in slate (killas) is also discouraging, but a soft nature both in granite and slate, and in the latter, the moderate thickness of the beds, and the slight inclination of the laminæ, are encouraging features. The veined and bedded structure of lodes, and their frequent curvatures are not inviting, neither are they rich when having a flat underlay. The quartzose, and generally speaking the smaller portions, are not so rich as those which consist of softer materials, and are of larger size. The occurrence of gozzan in the superficial parts, and the frequency of bunches of ore near cross veins are considered beneficial.

Even the most regular tin and copper lodes are very complex in their composition : quartz generally prevails in the matrix, but is always more or less blended with a substance similar to the adjoining rock ; indeed, the latter often occurs in distinct forms,—as nodules, angular pieces, and even masses of considerable size, which are independant of the main rock, being completely enveloped in the quartzose part of the lode. These are of such common occurrence, as to have been named by the miners, *horses of killas.* Sometimes the schist so abounds in the lode, that the quartzose part altogether disappears, or is only continued in minute strings ; in this case, the lode is said to have dwindled away, or to have been *wrung out.* It also frequently happens that both these principle parts (the rock and the quartz) are intimately united, producing a siliceous layer of rock which is still metalliferous, and is commonly called *capel :* hence the courses of schorl-rock, porphyry, and some anomalous rocks, which have been called by the miners *elvans*, have been properly considered by them to be analogous to lodes ; for they are, in fact, veins on a large scale. " By a true vein," Mr. Carne says, " I understand the mineral contents of a vertical or inclined fissure nearly straight, and of indefinite length and depth. Their contents are generally, but not always, different from the strata, or the rocks, which the vein intersects. True veins have usually regular walls,, (or the country standing against the vein, on each side) and sometimes a thin layer of clay, between the wall and the vein. Contemporaneous veins have been usually distinguished from true veins by their shortness, crookedness, and irregularity of size, as well as by the similarity of the constituent parts of the substances which they contain, to those of the adjoining rocks, with which they are generally so closely connected, as to appear a part of the same mass." Tin lodes are, in general, richer or poorer in the elvan, than in the adjoining rocks, in proportion to the hardness or softness of the elvan. A very soft, or very hard gossan (earthy brown iron ore) is equally thought less favourable than if its consistency be moderately firm ; and a very dark colour is also discouraging. The copper gossans are generally softer, paler, and less quartzose, or rather, perhaps, the quartz in them is often friable, and they are more vesicular than tin gossans.

In granite the lodes, which are chiefly productive of tin ore, are, for the most part, composed of a pale greenish felspar of a confusedly crystalline structure, but seldom containing distinct crystals. Through this substance the tin ore is interspersed in form of crystalline granules, seldom so large as a pea, but generally as small as sand.

The lodes which yield copper ore in granite almost always contain gossan near the surface, and this usually continues to somewhat greater depths than it does in slate,—as at Tresavean, Ting Tang, Dolcoath, &c., in Cornwall.

When the lodes are very granitic, or when they contain much of the schorlaceous quartz, they are seldom productive ; indeed, copper ores are rarely found embedded in schorl. The lodes which yield copper ores in slate contain large quantities of gossan of a pale hue, soft, and full of soft cavities. In them, also, tin ore frequently occurs in small quantities, and blende is very plentiful ; but iron pyrites (mundic) is almost constantly present. These earthy minerals are mostly quartz, mixed with quantities of felspar, clay, or fluccan ; near the surface these are spotted with earthy black copper ore, and at length by copper pyrites. In many places, and more especially in the slaty rocks in the neighbourhood of the fossilliferous beds in the eastern district of Cornwall, some portions of the lodes, when large, consist almost wholly of a very white chrystalline quartz abounding in vughs, or cavities lined with chrystals of the same, and enclose innumerable disjointed pieces of slate. The cavities lined with crystals, and the included spots of slate, are most unequivocal signs of poverty in those parts of the lodes where they occur. There are also certain minerals which are seldom found in the richer parts of lodes ; in those which yield copper ore, chlorite, provincially called *peach*, is one of the most conspicuous. The occurrence of tin ore in the deeper parts of lodes which have previously produced copper ore only, is accounted a very unfavourable indication. Ores of a certain character produce the same metal, and the miner, from experience, can immediately say which ore contains copper, which tin, and which lead.

It is generally, if not invariably the case, that a peculiarly favourable matrix for copper ore is found at the juncture of killas and granite, and the richest and most numerous veins are generally discovered in killas (clay slate) at no great distance from the granite, and are seldom sought after anywhere else by cautious miners.   The pale blue killas generally accompanies a rich vein of copper, and it is the easiest to work on, in sinking shafts and pursuing discoveries.   The lodes vary in width, from an inch to thirty feet, but the most general in tin and copper veins in Cornwall, is from one to three feet, and in the thinner veins the ore is less mixed with other substances.   A lode composed of beautiful spar, yellow ore, white iron, and a portion of mundic, is seldom known to fail making a great quantity of ore.   The *underlay* (or deflection from the perpendicular) of lodes is north and south.   If the north side of the roof of a church were, retaining its slanting position, supposed to be underground, it would give an idea of a lode. In deep mines the lode sometimes passes through the killas, and is continued in granite.

The fairest method of working a mine, and which is generally adapted in the best conducted, is to promote *discovery ;* ground being constantly opened, but not more than half the ore found to be taken away—the other half being left as a reserve, in case of any temporary falling off in the mine, that there may be something to fall back upon, whilst opertions are extended in search of more ; and great skill and judgement are required in a mining captain to arrange the workings, so as to keep up a regular and good supply of ores.

The legitimate value of a mine chiefly depends upon the value of the ore actually discovered underground, and the reasonable anticipations of further discoveries being made, as determined by the state of the mine and the richness of the district in which it is situated,—the value of the mineral produce in the market, and the value of the machinery, materials, and erections on the surface ; and persons entering upon mining with the view of a permanent investment would do well to remember this, and not to take as a sole criterion of the value of a mine, its having realized large profits ; for there is such a thing known to mines as " picking out the eyes" of a mine, or taking away the reserved ores, in order to make those very profits, and so raise a fictitious value for their shares in the market!

Many cases have occurred where every bunch of ore discovered has been exhausted, and the profits divided immediately ; so that, when the lode, for a time, became small and profitless, calls have been made upon the pockets of the proprietors for money to extend their operations, which should have been paid out of the produce ; and the mines, in consequence of not paying, have been " knocked" or abandoned by one party, and soon after taken up by another, who, by working fairly and properly, have made them both lasting and good.

THE SYSTEM OF CORNISH MINING.

THE management of some of the principal Cornish mines, is in the hands of a Committee, consisting generally of the largest shareholders in the county, or, as they are termed *in-adventurers* ; but, in the majority of mines, a gentleman chosen from the adventurers, and called the *purser*, has the entire management , keeping the accounts, and paying all monies. At the meetings of the adventurers, which are generally held on the mines *every two months*, the purser presents a statement of the accounts, when, if there be any profit, it is divided ; and, on the contrary, should the mine be in debt, calls are made upon the shareholders to defray them. At these meetings the best modes of working the mines are discussed, and carried by a majority of the shareholders present, according to the interests they hold. Next to the purser, is the head captain or manager, who superintends the whole of the mine, and the general routine of the surface work ; the underground captains, seeing that the work is there conducted properly ; the persons performing the work in the various parts of the mine, may be divided into tributers, tutworkmen, and labourers.

Tributers receive a certain portion of the ore, or so much in the pound, (as may be agreed upon,) in the value of what they raise. Tutworkmen work by the piece, generally calculated by the fathom ; in this way the shafts are sunk, adits driven, and the labour usually performed in those parts of the mine which do not produce ores ; the labourers are generally employed on the surface dressing ores, &c., and consist of men, boys, women, and girls. The general features of a mining district, has been graphically sketched by a talented writer.

" To one unaccustomed to a mining country, the view from Cairn Marth, which is a rocky eminence of seven hundred and fifty-seven feet, is full of novelty.   Over a surface, neither mountainous nor flat, but diversified from sea to sea by a constant series of low undulating hills and vales, the farmer and the miner seem to be occupying the country in something like the confusion of warfare.   The situations of the Consolidated Mines, the United Mines, the Poldice Mine, &c., &c., are marked out by spots a mile in length, by half a mile in breadth, covered with what are termed " the deads" of the mine—*i.e.* slaty poisonous rubbish, thrown up in rugged heaps, which, at a distance, give the place the appearance of an encampment of soldiers' tents.   This lifeless mass follows the course of the main lode (which, as has been said, generally runs east and west); and from it, in different directions, minor branches of the same barren rubbish diverge through the fertile country, like the streams of lava from a volcano.   The miner being obliged to have a shaft for air at every hundred yards, and the stannary laws allowing him freely to pursue his game, his hidden path is commonly to be traced by a series of heaps of " deads" which rise up among the green fields, and among the grazing cattle, like the workings of a mole.   Steam-engines, and *whims*, (large capstans worked by two or four horses,) are scattered about; and in the neighbourhood of the old, as well as of the new workings, are sprinkled, one by one, a number of small whitewashed miners' cottages, which, being neither on a road, nor near a road, wear, to the eye of the stranger, the appearance of having been dropped down *a-propos* to nothing.—Such, or not very dissimilar, is in most cases the superficial view of a country the chief wealth of which is subterraneous.

Early in the morning the scene becomes animated.   From the scattered cottages, as far as the eye can reach, men, women, and children of all ages begin to creep out ; and it is curious to observe them all converging like bees towards the small hole at which they are to enter their mine.   On their arrival, the women and children whose duty it is to dress or clean the ore, repair to the rough sheds under which they work, while the men, having stripped and put on their *underground* clothes, (which are coarse flannel dresses,) one after another descend the several shafts of the mine, by perpendicular ladders, to their respective levels or galleries*—one of which is nine hundred and ninety feet below the level of the ocean.   As soon as they have all disappeared, a most remarkable stillness prevails—scarcely a human being is to be seen.   The tall chimneys of the steam-engines emit no smoke ; and nothing is in motion but the great ' bobs' or levers of these gigantic machines, which,

---

* Recently, a vast improvement in the mode of descending deep mines has been accomplished, and is in successful operation at the Tresavean Mine.

slowly rising and falling, exert their power, either to lift the water or produce from the mine, or to stamp the ores ; and in the tranquillity of such a scene, it is curious to call to mind the busy occupations of the hidden thousands who are at work: to contrast the natural verdure of the country with the dead product of the mines, and to observe a few cattle ruminating on the surface of green sunny fields, while man is buried and toiling beneath them in darkness and seclusion.—But it is necessary that we should now descend from the heights of Cairn Marth, to take a nearer view of the mode of working the mine, and to give a skeleton plan of that simple operation."

A lode, as before stated. is a crack in the rock, bearing, in shape and dimensions, the character of the convulsion that formed it; and it is in this irregular crevice that nature has, most irregularly, deposited her mineral wealth ; for the crack, or lode, is never filled with ore, but that is distributed and scattered in veins and bunches, the rest of the lode being made of quartz, mundic, and ' deads.' Under such circumstances, it is impossible to say beforehand, where the riches of the lode exist : and, therefore, if its general character and appearance seem to authorise the expense, the mine is commenced in the manner before explained.

The object of perpendicular shafts, and horizontal galleries, is not so much to get at the ores which are directly procured from them, as to put the lode into a state capable of being worked by a number of men— in short to covert it into what may now be termed a mine. In the Cornish mines, the sinking of the shafts, and the driving of the levels, is paid by what is termed *tut-work*, or task work, that is, so much per fathom ; and, in addition to this, the miners receive a small per centage of the ores, in order to induce them to keep these as separate as possible from the *deads*, which they would not do, unless it were thus made their interest. The lode, when divided as above described, is open to the inspection of all the labouring miners in the country ; and by a most admirable system, each mass or compartment is let by public competition, for two months, to two or four miners, who may work it as they choose. These men undertake to break the ores, wheel them, raise them to the surface, or, as it is termed, ' *to grass*' and pay for the whole process of dressing the ores—which is bringing them to a state fit for market. The ores are sold every week by public auction, and the miner receives immediately the *tribute* or per centage for which he agreed to work—which varies from sixpence to thirteen shillings in the pound, according to the richness or poverty of the ores produced. The owners of the mine, or, as they are termed, the *adventurers*, thus avoid the necessity of overlooking the detail of so many operations, and it is evidently the interest of the miner to make them gain as much as possible. Should the *pitch*, or compartment, turn out bad, the miner has a right at any time to abandon his bargain, by paying a fine of twenty shillings. At the expiration of the lease, or whenever they may be abandoned, the *pitches* are anew put up to auction, and let for two more months : some may be getting richer, others poorer, as the work proceeds ;—and thus public competition practically determines, from time to time, the proper produce which the miner should receive. The different rectangular masses, or *pitches*, into which the lode is divided by the galleries and shafts, very seldom turn out to be of similar value ; and they are of course worked exactly in proportion to their produce. In one compartment the whole of the ore is worked out ; in another only a proportion will pay for working ; while not a few turn out so poor, that no one will undertake to work them at all. The *pitches* are in most cases taken by two miners, who relieve each other, and one often sees a father and son, who are in partnership, gradually find the lode turn out poorer and poorer, until they are at last compelled to pay their fine, and quit the ungrateful spot. The lottery in which the *tributers* engage abounds in blanks and in prizes. Sometimes the lode gets suddenly rich, sometimes as suddenly poor, and occasionally a productive lode altogether vanishes, or, as the miners say, has ' *taken a heave ;* by which they mean, that some convulsion of nature has broken the lode, and removed it off—sometimes two or three hundred feet— to the right or left. In order to determine where to find it, those well acquainted with the subject carefully observe the fracture or broken extremity of the lode, and from its appearance they can determine on which side, and in what direction, to search for the lost prize. Sometimes again, a lode which is paying very well, is all of a sudden found to have ' *taken horse*,' which means, that it has split into two lodes, separated from each other by an unproductive mass, which the miners term a ' *horse ;* and although the aggregate of the two lodes frequently contains the same quantity of ore as the original single lode, yet as the expense of working is doubled, it often will not pay to work them ; for in all mining operations it must be constantly remembered, that it is not the quantity, or even quality of the ores, that can induce a prudent man to work them, if the *expenses*, from any circumstances, should exceed the *returns*.

There is no light in a mine but that afforded by the candles of the workmen ; while the universal presence of water soaking through the crevices of the gallery, and intermixing with the dust and rubbish, keep up a constant succession of dirty puddles, rendering it no very pleasant affair going underground. Each miner has a candle, which is stuck close by him against the wall of his gallery, by means of a piece of clay ; and besides those employed in extending the gallery, there are generally one or two boys wheeling the broken ore &c., to the shaft. Each boy has a candle affixed to his wheelbarrow, by the universal subterranean candlestick, a piece of clay. The men relieve each other every six or eight hours, and thus keep on their work uninterruptedly, except on Sundays. Notwithstanding this incessant labour, the progress of the miner in excavating his gallery is, in general, very small; one, two, or three feet in a week, or a few inches daily, is often the whole amount of the united operations of twenty or thirty men. In loose lodes, and in killas districts, they cast more, but the lode is rarely so wide as the gallery or level, so that it becomes necessary to cut away the solid rock on each side, which is often very hard, even when the lode is soft.

In working by tribute, the miner naturally does all he can to enrich himself, but the system is so admirably balanced and arranged by long practice and experience, that it is very difficult for him to enrich himself without also enriching the owners or *adventurers*. Still, however, there are modes by which he occasionally endeavours to defraud his employer. The miners will sometimes steal each other's ores. If they come to a very good lode, they will occasionally hide their ore under the rubbish, or *dead*s, with the view of making the profit they are getting appear to be inconsiderable, and, of course, being able, at the end of their contract, to take on their *pitch*, for another two months, at an easy rate. They perhaps succeed in this; but when they go to reap the benefit of their fraud, they sometimes find that a brother miner, still more cunning than themselves, has discovered their hidden treasure, and has carried it off. The most usual mode of fraud, however, is a combination between two *tributers*, one of whom is working very rich, and the other very poor ores, The tributer who is working poor ores has, perhaps, bargained that he is to receive thirteen shillings out of every twenty shillings' worth of ore; while his friend, who is working the rich ores, is to get only one shilling out of twenty. In the dark chambers of the mine these two men secretly agree to exchange some of their ores, and then to divide the gross profits, which are, of course, very large; for, by this arrangement, instead of one shilling they get thirteen shillings out of twenty for a portion of the rich ores, while they lose but a trifle on a corresponding portion of the poor ores, There are a few other methods of defrauding the adventurers; but in the diamond-cut-diamond system of the Cornish mines, a severe check upon all such tricks is established in the appointment of a number of excellent men, who are selected from among the working miners, to superintend all their operations. These men, having been brought up in the mines, are, of course, acquainted with the whole system. They have fixed salaries of about eighty or ninety pounds a-year, and are termed *captains of the mines.*

The ores, or, as the miners term them, ' *hures*,' are all dressed by women and boys, who cob them, pick them, jig them, buck them, buddle them, and splay them as they may require ;—but as these terms of art may not be altogether intelligible, the process may be described in humbler words. In order to prepare *copper ores* for market, the first process is, of course, to throw aside the deads, or rubbish, with which they are unavoidably mixed ; and this operation is very cleverly performed by little girls of seven or eight years of age, who receive threepence or fourpence a-day. The largest fragments of ore are then *cobbed*, or broken into smaller pieces, by women ; and after being again picked, they are given to what the Cornish miners term ' *maidens*' —that is, to girls from sixteen to nineteen years of age. These maidens *buck* the ores,—that is, with a bucking iron, or flat hammer, they bruise them down to a size not exceeding the top of the finger ; and the *hures* are then given to boys, who *jig* them, or shake them in a sieve under water, by which means the ore or heavy part, keeps at the bottom, while the spar, or refuse, is scraped from the top. The part which passes through the sieve is also stirred about in water, the lighter part is thrown from the surface, and the ores, thus dressed, being put into large heaps of about a hundred tons each, are ready for the market. They then are forthwith shipped for *Wales*, (it being much cheaper to carry the ores to the coals than the coals to the ores); and in Wales, after undergoing another trifling operation, they are ready to be smelted—a process of which no Cornish copper-miner of any order has the slightest notion, but which will be noticed hereafter.

The dressing of *tin ores* is altogether a different process, because not only are the ores perfectly different, but the method of smelting them is also so different, that it is necessary the tin should be reduced to the finest powder, while copper ore is smelted in small lumps. The tin ore, after being picked, or separated from the *deads*, is thrown into a stamping mill, where it gradually falls under a number of piles or beams of wood, shod with iron, which are worked vertically up or down—generally by a water-wheel. As it is necessary that the ore should be bruised to a very fine powder, the bottom of the stamp is surrounded by a very fine copper sieve, and water being made constantly to flow through this, the ore can only escape when it is fine enough to pass with the water through the interstices of the sieve. It then settles into a very fine mud, which is composed of metallic particles, and powdered quartz-rock, &c. This mud undergoes a very ingenious process, which the miners term *buddling*. The metallic and other particles are all of different specific gravities, and the dresser, being aware of this, places the mud at the top of an inclined plane, and, gently working it about, allows a small stream of water to run over it. In a short time the inclined plane is all equally covered with the mud, and although, to any person who has not been brought up to the business, the whole mass has the same appearance, yet the dresser is able to distinguish, and to draw a line between, the heavy metallic particles, which have remained at the top of the inclined plane, and the worthless ones which, from being lighter, have been washed towards the bottom. After separating the one from the other, the worthless part is thrown away and the metallic part buddled again, and the process is repeated until the mass retained consists almost entirely of metallic particles. But these particles, which are as fine as flour, are not all tin ; generally many of them are composed of mundic (the sulphuret of arsenic) ; others are copper ; and as the difference between the specific gravities of these three metals is not sufficient to separate them by buddling or washing, it becomes necessary to roast the mass, an operation which the dresser does not himself perform. As soon as the mass is placed in a furnace, and subjected to a proper degree of heat, the sulphuret of arsenic goes off in white poisonous fumes or smoke, and the specific gravities of the different particles of copper and tin are so altered by the action of the fire, that, upon being taken out of the furnace, and again delivered to the dresser, he finds that, in the course of carefully buddling the mass on the inclined plane before described, the particles separate—the tin, which is the heaviest, being left upon the upper part, while the copper is at the bottom. The tin is then packed in bags and sold ; and, being nearly pure metal, it requires, in comparison to copper ore, so little fuel, that it is all smelted *in Cornwall.*

Whoever compares together the two processes of dressing copper and tin ores, must be satisfied that they are completely different affairs ; and in Cornwall, accordingly, it is perfectly well understood that they form different trades. The ores are so dissimilar, and require such different modes of treatment, that the experience which the labourer gains in dressing the one, is of no possible use to him who dresses the other. It is true that both sets of people are called *dressers,* but it does not follow that, for that reason, they can all dress *any* thing ; and to desire a copper-dresser to dress tin ores would, in Cornwall, be considered as preposterous as if one were to send him to Aldersgate-street to dress a turtle, or to St. James's-square to dress a duchess.

But it is time that the *underground captains* should come *to grass,* and that the whole body of subterraneous labourers should be released ; and those who have attended to their labours throughout the day, will scarcely regret to see them rising out of the earth, and issuing in crowds from the differents holes or shafts around—hot—dirty—and jaded ; each with the remainder of his bunch of candles hanging at the bottom of his flannel garb.   As soon as the men come *to grass* they repair to the engine-house, where they generally leave their *underground clothes* to dry, wash themselves in the warm water of the engine-pool, and put on their clothes, which are always exceedingly decent.   By this time the *maidens* and little boys have also washed their faces, and the whole party migrate across the fields in groups, and in different directions, to their respective homes.   Generally speaking, they now look so clean and fresh, and seem so happy, that one would scarcely fancy they had worked all day in darkness and confinement. The old men, however, tired with their work, and sick of the follies and vagaries of the outside and the inside of this mining world, plod their way in sober silence—probably thinking of their supper.   The younger men proceed talking and laughing, and where the grass is good they will sometimes stop and wrestle.   The big boys generally advance by playing at leap-frog ; little urchins run on before to gain time to stand upon their heads ; while the " *maidens*," sometimes pleased and sometimes offended with what happens, smile or scream as circumstances may require.   As the different members of the group approach their respective cottages, their numbers of course diminish, and the individual who lives farthest from the mines, like the solitary survivor of a large family, performs the last few yards of his journey by himself. On arriving at home, the first employment is to wheel a small cask in a light barrow, for water—and as the cottages are built to follow the fortunes and progress of the mine, it often happens that the miner has three miles to go ere he can fill his cask.   As soon as the young men have supped, they generally dress themselves in their *holiday-clothes*, —a suit better than the *working-clothes* in which they walk to the mines, but not so good as their *Sunday-clothes*.   In fact, the *holiday-clothes* are the *Sunday-clothes* of last year, and thus, including his *underground flannels*, every Cornish miner'generally possesses four suits of clothes.

The Sunday is kept with great attention.   The mining community, male and female, are remarkably well-dressed ; and as they come from the church, or meetings, there is certainly no labouring class in England at all equal to them in appearance, for they are naturally good-looking. Working away from sun and wind, their complexions are never weather-beaten, and often ruddy; they are naturally a cheerful people, and, indeed, when one considers how many hours they pass in subterraneous darkness, it is not surprising that they should look upon the sunshine of the Sabbath as the signal, not only of rest, but of high and active natural enjoyment.

The ticketing, or weekly sale of the ores, forms a curious feature in the system of Cornish mining.

The copper ore, on being raised from the mines and dressed, is put into heaps of several tons ; one is well mixed, and any sampler on an appointed day fixes on a third or fourth of the dole. After sub-dividing and mixing this, a sufficient quantity is put into a bag by each sampler, and this is the sample of the whole. These are carried to the different assay offices, where the ore is pulverised, and an ounce, troy, assayed in a crucible, with proper fluxes, and a bead, or prill of copper, is found among the scoria, If an ounce of ore, yield one dwt. of copper, the produce of that ore will be one in twenty, or five per cent., and so on.

The *standard* of copper is the term given by the smelter to denote the price of a ton of metal in the ore, from which standard he deducts £2. 15s. for every ton of ore, or as many as may be required, according to its produce, to give a ton of copper ; and which sum is considered by the smelter as an equivalent for the returning charge, or expense of reducing the ore to a merchantable state. Suppose a parcel of ore makes a produce of eight and one-eighth per cent. at a standard of one hundred and twenty (that is, the price which the purchaser can obtain for a ton of metal ;) the price of a ton of that ore may be thus obtained :—

$$
\begin{array}{l}
120 \\
8 \ \text{⅛ produce.}
\end{array}
$$

$$
\begin{array}{l}
\text{⅛}—960 \\
\phantom{\text{⅛}—}15
\end{array}
$$

100⦃ 9,75 ⦄£9.
    900

75
20

100⦃ 15,00 ⦄15s.

£9. 15s.
2. 15s. returning charge.

£7. 0s. value of the ore per ton.

Every twenty shillings the standard rises or falls will make a differ-ence in the assay of one shilling, or a twentieth in every pennyweight, and a halfpenny in every grain ; as, for instance, 1 dwt. 1 gr. at £95. standard, will make the produce £4. 15s. the dwt., and 3s. 11½d. the grain ; but if the standard be £96. the produce will be £4. 16s. the dwt., and 4s. the grain, deducting for returning charge.

A fortnights' interval takes place between the assay and the ticket-ing, during which time the agents receive answers from their principals, as to the price to be offered. Before dinner, tickets containing offers from the different copper companies founded on these assays are pro-duced, and the highest is the purchaser.

The Cornish assayers, generally, have not the slightest notion of the theories of chemistry or metallurgy, and their assays are not very accurate. Ore, which according to their assay gives a produce of seven, will often, upon a stricter analysis, yield more, and the difference goes to swell the already enormous profits of the smelter.

The process is, to take four hundred grains of the sample of ore, pound it fine, sift it, and place it in a crucible to roast in an air furnace, keeping the ore stirred with an iron rod. When the sulphur is considered to have been sufficiently driven off, the ore is taken from the fire, and allowed to cool gradually in the crucible ; if, then, the upper part appears red or brown, and the under part black, the proper roasting is supposed to have been given. A standard flux—composed of borax five dwts., lime one and-a-half ladleful, (diameter of the ladle about three-quarters of an inch, and depth half an inch,) and powdered fluor spar one ladle—is then mixed with the roasted ore and put into a crucible, the mixture being covered with salt. It is then melted, and what is termed a *regal*, or *regule*, produced ; this regule is thought good, if it will produce from eight to twelve in twenty. The grey sulphurets, the black oxides, and the carbonates, have sulphur added to them to " throw back the ores," as it is termed, as they are considered not to have enough of it for the purposes of the assay. To fine this regule it is pounded and roasted in a crucible until the sulphur is considered to be driven off. A flux—of nitre three dwts., red tartar ten dwts., borax five dwts., and salt two ladles—is then added, and salt sprinkled over the top of the mixture. Coarse copper is now obtained. If this comes out clean, the assay is put into a crucible without flux, and when melted, the crucible is taken out of the furnace, and shaken until the surface appears blue. A refining flux is now prepared, by mixing two parts nitre and one part of white tartar in an iron mortar, and stirring the mixture with a red-hot iron until deflagration has ceased. The flux thus prepared is powdered and sifted when cold, and five dwts. of it are added to a ladleful of salt, put with the assay into a crucible. When all is melted, the copper is poured into one mould and the slag into another ; the latter is again melted with two ladles of red tartar, and the small button, or prill of copper, now found, is added to that previously obtained, and the assay is completed. To assay tin in the dry way, reduce a portion of the ore to powder, place it in a crucible, or on the slab of a muffle, and expose it to a low red heat ; if arsenic or sulphur be contained, it will thus be got rid of. The residue, when mixed with a little charcoal and linseed oil, in a well closed crucible, is to be subjected to a bright red heat, by which it will be reduced to the metallic state.

The process of smelting copper ore is too complicated and lengthy for a small work like the present, and has but little interest, except to the practical miner ; however, the following very condensed account may give the reader some little idea of it :—Copper ores contain sulphur, iron, and arsenic ; the sulphur and arsenic being gradually dissipated in the furnaces. The first process is to calcine the ore, the heat being as great as the ore will bear without being fused or baked. When sufficiently cool to be removed, water is thrown over it to prevent the escape of the finer particles. The calcined ore is next melted, when the earthy matter, and metallic oxides, being specifically lighter, float on the surface, and are skimmed off. The metal is made to flow into a pit of water, where it becomes granulated and called *coarse metal*. If the slags contain any copper, on being broken, it is found at the bottom. The granulated metal contains about one-third of copper, composed chiefly of copper, iron, and sulphur. The coarse metal is next calcined, and after calcination, melted ; with the calcined metal are melted some slags, from the last operation, which contain some oxide of copper, which becomes reduced by a portion of the sulphur, which combines with the oxygen and ;passes off as sulphureous acid gas, while the reduced metal combines with the sulphuret. The slags being composed chiefly of the black oxide of iron, fuse readily, and act as solvents for earthy matter, &c. The metal after the slag is skimmed off is either tapped into water, or into sand-beds In the granulated state it is called *fine metal ;* in the solid form, *blue metal*, from the colour of its surface. The fine metal is next calcined and then melted, similar to that of the coarse metal, and the product is a coarse copper, containing from 80 to 90 per cent of pure metal. The next process is roasting or oxidizing. The pigs of coarse copper from the last process, are exposed to the action of the air, which draws through the furnace at a great heat. The volatile substances are expelled, and the iron or other metals which remain are oxidized. The pigs are covered with black blisters, and the copper is called *blistered copper*. It is porous and honey-combed, from the gas formed during the ebullition which takes place in the sand-beds or tapping. An assay is next taken out with a small ladle, and broken in a vice ; the copper in this state is termed *dry*. In the process of toughening, the surface of the metal in the furnace is first well covered with charcoal ; a pole, commonly of birch, is then held in the liquid metal which causes considerable ebullition, owing to the evolution of gaseous matter ; and this operation of *poling* is continued, occasionally adding fresh charcoal, that the surface may be covered until the refiner perceives, by repeated assays, the grain perfectly closed, so as to assume a silky polished appearance when half cut through and broken, and is become of a bright red colour. If it be left under the hammer, and do not crack at the edges, he is satisfied of its *malleability.* Copper for brass is granulated with warm water, the copper assumes a round form, and is called *bean shot.* With cold water it has a light ragged appearance, and is called *feathered shot.* The former is the state in which it is prepared for brass-wire making. It is cast for exports to the East Indies in pieces six inches long, and weighing eight ounces, called *Japan copper.* These are dropped from the moulds, immediately on becoming solid, into cold water, and by a slight oxidation, the copper acquires a rich red colour on the surface.

Tin ore, previous to its reduction, is pulverised in a stamping mill, to various degrees of fineness, depending on the size of its crystallisation, and the ingredients with which it is found mixed. If it contain no volatile matter, it is seldom desiccated ; on the contrary, if it does, this process is never dispensed with. The ore is reduced in a reverberatory furnace, mixed with a portion of culm. At the east end of the furnace is a *ridge,* or railed line across the floor, between which and the eastern wall is the fire ; the chimney is at the west end, and its mouth about two feet above the level of the ore, so that the heat passes over the whole most intensely ; as it melts, the combustible and earthy matter, being specifically lighter, floats on the surface. By introducing an iron rake, the brilliant metal momentarily meets the eye, below the buoyant covering. In about six hours the metal is let out, and the floor remains covered with scoria. This matter is drawn out at the west end ; and when cooled is black from the prevailing oxide of iron. The mode of purifying tin from its alloys, differs little from that observed in refining copper. The tin smelted at the different houses is cast into moulds, containing about 3 cwt., and while in a fluid state it receives the stamp of the particular house, where it is smelted, and is denominated *block tin.* The blocks are weighed, numbered, and sent to the nearest coinage-town to be coined. In the coinage hall, a piece of about three or four ounces is cut off from one of the lowest corners, in order to prove the fineness of the metal. The face of the block is then stamped with the duchy seal, which constitutes the coinage, and is a permit for the owner to sell ; and, at the same time, the corner being cut off is an assurance that the tin has been properly examined, and merchantable.

The process of reducing gold, silver, lead, zinc, &c., will be found under their respective heads in " the History of Metals."

The quantity of timber used annually in the Cornish and Devon mines is very considerable, amounting to about £50,000., and consists almost entirely of Norwegian Pine, upon the importation of which a drawback is allowed.

The discovery of gunpowder forms a grand epoch in the history of mining ; but it is difficult to ascertain the exact time when *blasting* first came into use among Cornish miners. It was first used in Hungary or Germany, about 1620, and was introduced into England at the copper mines at Ecton, in Staffordshire, by German miners, brought over by Prince Rupert. It was not known in Somersetshire until 1634, after which the Cornish became acquainted with it; and it is supposed to have been first used in the district of Leland, Zennor, and St. Ives, by two men who came from the East, named Bell and Case, and who kept their operations a secret, suffering no one to see them charge the holes, till a man of Zennor, hiding himself upon a bolt, saw what they were about. In blasting, a hole is made in the rock with a steel borer, which hole is filled with gunpowder, the force being confined with a wedge, or by " tamping" over with some soft material, and is then set fire to, by means of a safety fuse, lighted at some distance, and large portions of the rock or lode, are forced off. The annual value of gunpowder used in Cornish mines has been estimated at £13,200, the quantity being about three hundred tons of two thousand pounds each.

The steam power employed in Great Britain, for mining purposes, may be estimated as amounting to the labour of one hundred and fifty thousand horses, or to that of seven hundred and fifty thousand men. The first steam engine erected in Cornwall, (under the plan of New-comen, who obtained a patent in 1705) was on Wheal Vor Mine, in Breage, between the years 1710 and 1714 ; the second at Wheal For-tune, in Ludgvan, in the year 1720. Newcomen's were superseded by Watt's engine in 1778 ; one of the latter, of thirty inch cylinder, being then at work at Wheal Busy (Chacewater.) Pryce describes this en-gine as working " a pump of six-and-a-half inches in diameter, in two shafts, by flat rods, with great friction, three hundred feet distant from each other, forty-five fathoms high on each shaft, equal in all to ninety fathoms, ' and as making' fourteen strokes of eight feet long per minute, with a consumption of coals less than twenty bushels in twenty-four hours." To ascertain the amount of fuel saved by working Watt's engines, a *counter* was invented, which, being attached to the main beam, marked the number of its vibrations, from which the work done by the engine was calculated, and the amount of coal consumed being ascertained, the saving was found. In 1812, Captain Joel Lean sug-gested the plan of placing a *counter* on every engine in Cornwall, and of publishing the duty performed by them; which is estimated by as-certaining the number of pounds' weight which are lifted one foot high by them, by the consumption of one bushel of coals. Capt. Lean, by most of the mining adventurers, is appointed to fix a counter upon the various engines to be reported monthly ; the counter is furnished with a Bramah's lock, the key of which Capt. Lean keeps, and it is inspected by him once per month. Sometimes another counter is attached to the engine, which is open to the inspection of the engineer, the agents of the mine, and the engine-men. A separate party supplies the coals which are delivered to the order of the engine-men, as they may require them, the orders being first examined and countersigned by Capt. Lean. At the end of each month, the coals not consumed are measured, and thus the real amount consumed is ascertained. This being done, and the counter examined, the *duty papers* are published, and include not only an account of pumping engines, but also of those employed in drawing ores up the shafts, and in stamping ores.

The Cornish pumping engines of the present day stand pre-eminent, and mines are worked which must long since have been abandoned but for them.

In large adventures the ores are very commonly raised to the surface by steam-whims, one of which is contrived, by the means of flat rods, to draw from two shafts, and sometimes three ; and these engines afford great advantages in working a mine where water is scarce, and horse-whims would be insufficient. In the eastern district, where water is more plentiful, there are several large water-whims, especially at Fowey Consols, and Wheal Friendship, (which are described under the heads of those mines). Where the supply of water is precious, large sums of money are often paid in Cornwall and Devon for the use of it ; and no small contrivance is frequently exhibited in turning a stream to the greatest account, which will be seen on referring to Wheal Uny Mine, in the Gwennap district. The Charlestown mines in St. Austle, pay £350. a year water rent.

The principal operations in mining having been noticed, a description of the different districts, with a statistical account of the principal mines in each, follows next in order.

THE MINING DISTRICT OF GWENNAP.

INCLUDES portions of the parishes of Redruth, Gwennap, Perranwor-thal, Kea, and Kenwyn; its rocks are the north-eastern skirt of the great granitic range of Stythians, Wendron &c.; and the mass of granite forming Cairn Marth, and Trefula Beacon, and the slate rocks in contact with them; the whole is traversed by numerous elvan courses, lodes, cross-courses and fluccans, in cross veins composed of clay; and is by far the most extensive as well as the most productive mining district in Cornwall, particularly in copper ores.

The mineral composition of the granite is much the same as in the other districts, viz., a basis of felspar, quartz, and mica, enclosing crystals of felspar, and occasionally quartz. In many places schorl abounds, At Tresavean, a schorl rock intervenes between the granite and the ordinary slate of the district; and in Wheal Beauchamp also, a similar rock prevails. Whenever the nature of the surface will permit, varieties of slate are seen in the Gwennap moors, and near Trevarth and Comfort a greyish buff, and blue-coloured slate passes through Ale-and-Cakes, the eastern part of the Consolidated Mines, and the Union Mines, but in all cases it assumes a darker hue at greater depths, and, generally speaking, the lodes are less productive in this deep blue slate, than when lighter tints prevail. The cleavage dips towards the south-east, with but few exceptions.

The elvan courses are very numerous in this district; the most northerly one is on the north of Wheal Peevor, and may be traced across the common as far east as North Briggan, where it has been quarried; it dips north, and is from six to eight fathoms wide. Another elvan has been extensively worked in several places at Tre-leigh, north of Redruth, passing through Cardrew-downs and Tres-kerby, it is from eight to ten fathoms wide, and dips south; at Cardrew it sends off shoots into the slate.

There is an elvan also in the adit at Wheal Buller, and again seen at the seventy fathom level; another in Beauchamp engine shaft, at a depth of thirty or forty fathoms it is only three fathoms wide, whilst at seventy fathoms deep, its breadth is thirty fathoms. The great elvan rising north of the valley leading in to Unity-wood, has been extensively worked for tin ore, which is abundantly sprinkled through it. The next elvan course passes through Wheal Unity and Poldice, and is intersected by many of the workings of Creeg Braws.

The elvan courses abound in irregular joints; they are commonly several fathoms wide, and their composition, veins of compact felspar, usually of a greyish-buff or pale pink hue, mixed with some silicious matter, and generally containing masses of crystalline felspar, and often too, double pointed crystals of quartz, schorl not unfrequently appears in radiating crystalline groups, and ferruginous matter is very largely dispersed through them. As the mines long since abandoned will not be placed in their respective districts, it may be as well perhaps, to name the principal of them here ; with the profits made by each. Wha Abraham, Crenver and Oatfield, profits £150,000. Wheal Bassett, £100,000. Herland, £90,000. Wheal Chance, £150,000. Wheal Music, £100,000. Wheal Spinster, £80,000. Wheel Speedwell, £60,000. Treskirby, £200,000. Union Mines, £50,000. Camborne Vean, £200,000. Binner Downs, £100,000. Stray Park £40,000. (Resumed, and likely to do well.) Wheal Trannack, £20,000. Wheal Trenwith, £40,000. Wheal Tolgus, £50,000. Wheal Leisme, £110,000. Great Wheal Towan, £250,000.

It may be as well to observe, that the statistical information contained in the following pages has been compiled from official documents and reports.

THE CONSOLIDATED COPPER MINES.

In the parish of Gwennap, consisting of Carharrack, West Wheal Virgin, Wheal Virgin and Wheal Fortune, are the largest in Cornwall, the workings extending 63 miles underground, or 55,000 fathoms ; and they have made a profit of upwards of £700,000. Previous to consolidation, they returned 17,000 tons of ore per annum, and left a profit of £400,000. and Wheal Virgin, now a part of these mines, in July and August 1757, produced in one fortnight, ores which sold for £5,700., and in the next three weeks and two days, as much as sold for £9,600, The cost of raising the former being only £100, and that of the latter a trifle more, in proportion to the quantity. In 1819, a company was formed to work these mines under the management of Mr. John Taylor. F.R.S., but the lease having expired in 1840, they are now under the management of the Messrs. Williams, of Scorrier House. During the management of Mr. Taylor, the mines returned upwards of 300,000 tons of ore, yielding £2,000,000. The cost for working them amounted in the twenty-one years to £1,500,000., and about £300,000. profit was divided among the adventurers. The present proprietors commenced working in July, 1840, and from that time to June, 1842, have returned 25,807 tons of ore, yielding £161,120. 1s. 6d., out of which they divided a profit of £16,500. up to March 1842, but since that time, the mines have not paid expenses. They have eight large steam-engines, and about thirty small ones at work. The mine is 320 fathoms deep from the surface, 2,000 persons employed, and the cost of working, averages about £5,000. a month. The mine is held in 100 shares, and the machinery valued at £70,000.

TRESAVEAN COPPER MINE

IN Gwennap, was once or twice abandoned as a failure ; but at length taken up by a party who persevered in exploring it, and with an outlay of little more than £1000, succeeded in discovering its wealth ; and its continued richness offers an extraordinary instance of fortunate adventure. The mine is a very dry one, situate on the slope of a hill, and requiring comparatively little machinery to draw the water from it ; the lodes are principally in granite, becoming profitless when they quit it and pass into the slate.

Tresavean has left a profit of more than £800,000 ; the profits for the last five years have averaged £30,693 a year; the working cost being between 3 and £4000 a month. There are 1,300 persons employed, and the machinery is valued at £60,000. During the last eight years from June, 1834, to June, 1842, there have been raised and sold from the mine 99,211 tons of ore yielding £610,893 19s. 6d. The mine is held in ninety-six shares.

They have several steam engines at work, and the new engine shaft lately completed is two hundred and seventy-six fathoms deep from the surface, and took two years and two months sinking, by twelve sets of men rising and twelve sets sinking ; in all, one hundred and twenty men at the same time employed. On this shaft a steam engine, with a cylinder of eighty-six inches diameter, has been erected, which works nine lifts of pumps, and lifts thirty six tons six cwt. per stroke ; the weight of rods and setts off in the shaft is fifty-nine tons thirteen cwt. two qrs. ; the shaft main beam, with gudgeons, bearers, and connection, fifty tons ; eight plungers, seven and-a-half tons ; four balance bobs, sixty tons ; four balance boxes, eighty tons ; seventy-five fathoms of flat rods underground, eleven and-a-half tons. Total weight of engine when in motion, three hundred and fifty-three tons sixteen cwt. The price of this engine delivered on the mine was £4,185. The size of the shaft is twelve feet by six, and cost upwards of £20,000 sinking.

A machine for raising and lowering the miners has lately been completed in this mine. It is the invention of Capt. Michael Loam, and is formed of two perpendicular rods of wood, having projections about twelve feet apart, upon which each man, ascending or descending, stands. In the rods are placed long iron handles, which the men lay hold of with the greatest ease. As one rod descends, the other ascends, and at every alternate step, there is a slight check, which affords sufficient time to enable the person travelling to remove from one rod to the other. The movement of these rods enable a man to travel about eleven fathoms a minute. The machine is carried to a depth of one hundred and forty fathoms, and worked by a thirty-six inch double rotatory engine, acting upon two small wheels, which act upon two larger ones. A model of this machine may be seen in the Polytechnic Institution.

PENSTRUTHAL COPPER MINE

IN Gwennap, at a former working, left a profit of near £100,000. The mine is not more than seventy fathoms deep. In fact, most of her riches were found above the sixty fathom level ; and during three years, from 1826 to 1829, ores were raised to the value of £132,186. , but in consequence of this extravagant working, the ore failed, and the mine was abandoned about six years ago. It has, however, been lately resumed in one hundred and twenty eight shares, and with good manage-ment, likely to do well, as they have a fine course of ore.

THE UNITED COPPER MINES

IN Gwennap, consist of the mines of Poldory, Cupboard, and Ale-and-Cakes ; and, in the first working, divided a profit of £300,000. ; but, through some cause or other, gradually fell into decay, and caused a loss, eventually, of £50,000. After a time they were resumed, with an outlay and loss of £30,000. They then became the property of the Consolidated Mines' proprietors ; but the latter mines changing hands, in 1840, the United are now worked by a separate company, and, in twelve months, ending June, 1842, made a clear profit of £10,699. 10s., the cost for working during that time having been £53,450. 8s. 3d., and the produce £64,149. 9s. 1d. From June, 1841, to June, 1842, the produce has been 10,195 tons of copper ore, yielding £64,377 15s. 6d., but which, however, has not much more than paid the expenses of working. The proprietors are 100 in number, and the mines under the management of John Taylor, Esq., F.R.S. Near 1,500 persons are employed. The deepest level is 200 fathoms ; the engines are nearly at the extent of their power, and a new one is required, for which a new shaft must be sunk.

Beautiful specimens of arseniate of lead have been found in these mines. And the old copper lode has been traced from part of Camborne, through Carn Kye, to the mine of Baldue, a distance of seven miles.

WHEAL JEWEL COPPER MINE

In Gwennap; at the first working, left a profit of £200,000 ; the mine was then idle for twenty years, and was resumed about nine years ago by the Messrs. Williams of Scorrier House ; and during this short time has realised profits of upwards of £100,000. The sett is not more than 200 fathoms in length, and 150 in width, and the workings are principally in granite ; they have one steam-engine and two steam-whims ; the adit in Wearne's shaft is 50 fathoms from the surface, and the shaft 150 fathoms below the adit. About 300 persons are employed ; and from 1834 to June, 1842, have returned 31,276 tons of copper ore, yielding £214,130. 6s. 6d.

UNITY WOOD TIN AND COPPER MINE

In Gwennap, and Kenwyn ; at a former working, left a profit of £50,000, and was then abandoned through poverty. After this, the mine was taken by another party, who discovered rich courses of tin and copper, under the old workings, and for some time made profits to the amount of £14,000 a-year. This mine, at present, is worked with very little spirit, and the lease of the present company has nearly expired. About 200 persons are employed ; and in four years, ending June 1842, have returned 8,538 tons of copper ore, yielding £49,051.

POLDICE AND WHEAL UNITY MINES

In St. Day ; at a former working, Poldice left a profit of £150,000., and Wheal Unity £300,000, and since they have been consolidated have made small profits ; they are 250 fathoms deep, employ 300 persons ; and in four years, ending June, 1842, have returned 6,307 tons of copper ore, yielding £44,427. 11s. 6d.

TRETHILLAN COPPER MINE

In Gwennap ; adjoins Tresavean to the west, and is on the same lodes. The mine is worked through Tresavean shaft, and by the aid of her machinery, consequently at little cost (not more than £600. a month.) The sett is small, being about 84 fathoms in length, on the course of the lodes ; and the deepest level is 170 fathoms from the surface. This mine first made returns in April, 1837, and from that time to November, 1842, has returned 17,049 tons of ore, yielding £70,693., out of which the proprietors (120 in number) have divided a profit of £27,000, and are now receiving at the rate of £7,000. a-year.

WEST TRETHELLAN COPPER MINE.

In Gwennap; lies to the west of Trethellan, and to the south of Brewer, and the lode from the latter it is expected will run seventy or eighty fathoms in this sett. Very little has been done at present.

BREWER COPPER MINE

In Gwennap ; lies to the west of Trethellan and Tresavean, and is on the same lodes ; the sett is two hundred and sixty fathoms long, and Tresavean lode is supposed to run through it for more than one hundred fathoms. The mine has been at work two years, is seventy-five fathoms, deep, and returned, down to October 1842.—303 tons of ore, yielding £1,618, 2s. and leaving a profit of more than £700., the monthly cost not being £100.

TREVISKEY AND BARRIER COPPER MINES.

In Gwennap ; to the east of Tresavean Mine, the rich lodes of Tresavean in the deeper levels, are driven to the Barrier, which is a piece of ground around Tresavean Sett, five fathoms in width, and on the east, adjoining Treviskey, is four and a-half to five fathoms in length on the course of the lode. This sett, it is to be hoped, will be worked through Tresavean by the aid of her shafts and machinery as, by that means, large and immediate returns may be made, whereas, by sinking from the surface, it will be some years before the rich ore ground is reached, and that, at an outlay to the proprietors (120 in number) of from 40 to £50,000.

BELL COPPER MINE.

In Gwennap ; was formerly called Little Bell, or Bell Vean. The sett is three to four hundred fathoms long, on Penstruthal Lode, bounded to the west by Penstruthal, and east by Comfort Mine, and on the Tresavean North Lode; the present company have lately commenced working, and have expended £3,936.

WHEAL COMFORT COPPER MINE

In Gwennap ; formerly called Bell Veor, is situated north, and adjoining Tresavean Mine, and on Penstruthal South lodes ; they have lately commenced working, and have cut ore on the Tresavean North lode, and have returned down to Oct. 1842, 35 tons, yielding £210. 7s. The expenditure has been more than £2,000.

GRAMBLER AND ST. AUBYN UNITED MINES

IN Gwennap; were worked many years ago, and commenced again by Capt. W. Mitchell in 1834, who sunk forty-seven fathoms to the adit, and drove forty fathoms to intersect the shaft, when, (owing to Captain Mitchell's death) the mines were abandoned, after an outlay of £2,580. Mr. Adam Murray, and Capt. Thomas Kitto, jun., then became the purchasers, and the present company commenced working in 1840. There are several lodes, both of tin and copper, in the sett (which is surrounded by good mines,) and held by lease for sixteen years, at one-fifteenth dues. To the present time, near £10,000. have been expended by the proprietors, two hundred and forty-four in number, and have returned ores to the amount of £618.

WHEAL BUSY

IN Kenwyn ; formerly called " Chacewater," is the oldest copper mine in Cornwall, and was working to a profit in 1718, and during the first working left a profit of £200,000. From 1814 to 1820 it was, as the Chacewater Mine, returning ores to the amount of £20,000 a year ; but falling into decay, it changed hands, and is now worked as Wheal Busy, the operations being principally carried on above the adit and returning ore to the amount of about £1,200 a year. The mine is about two hundred and twenty fathoms deep from surface.

CREEG BRAWS MINE

IN Kenwyn ; lies to the north of the Consolidated Mines, and on the same lodes as Wheal Unity. This mine was very extensively worked, as far back, it is supposed, as the arrival of the Phœnicians in Cornwall,—the old pits, and other surface appearances, seeming to warrant such a presumption ; but there are no records extant of an earlier period than 1720, when it was worked by a Captain Richard Williams, under whose management considerable profits were made, and as much as five hundred thousand sacks of tin returned annually. There was also sold some copper ore of good produce, but the formations of that mineral were very " bunchy," and consequently the returns irregular. The water was pumped out of the mine by means of horizontal rods, connected with a steam engine working at Wheal Busy, and a water wheel was erected under ground for the same purpose, and propelled by water taken from Bizza Pool Mine. On the death of Captain Williams he was succeeded by his son, under whose management, however, the returns greatly diminished. The late John Williams, Esq., of Scorrier

House, now became interested, and the old mine was nearly abandoned, the operations being confined to the western part of the sett : about Gill's, Lake's, Taylder's, and Pollard's shafts, very little explorations were made for tin, although small quantities were occasionally raised about Pollard's shaft.   After this a Captain Paul and Captain Alex. Bray were appointed managers, and extracted sufficient ore to realize profit, but not to any considerable amount.   The management next devolved upon Mr. Joseph Morcom, of Whitehall, who raised large quantities of ore ; but, in consequence of the suppression of the tutwork, or exploring department, the returns were unequal to the expenditure, and the mine was abandoned.   From 1814 to 1822 the returns were two thousand five hundred and forty-two tons of copper ore, yielding £13,274.

In 1822 Mr. Williams obtained new grants for tin and copper at one-twelfth dues ; but from that period, little was done till the present adventurers took possession in 1840.   The mine is now under the management of Captain Lean and Mr. F. Blamey Purser, and likely to make a profitable one.   The largest proprietor is Mr. Robert Freeman, of Swanton Morley, Norfolk.   The cost expended by the present company is about £7,000, and the returns nine hundred and sixty-nine tons of copper, yielding £7756 7s.

WHEAL UNY TIN AND COPPER MINE

In Redruth ; formerly one of the " Redruth United Mines," is now worked by Messrs. Fox and Co., and returning small quantities of ore, but not yet paying expences.   They have a large steam engine pumping water to work water wheels for breaking and pulverizing the tin stuff ; the water is drawn by the steam engine from the bottom to the top of Wheal Uny Hill, and forms two separate mill streams, and after turning all the wheels in its course, returns again to the engine shaft through an adit at the bottom of the hill, and is again drawn up by the engine for the same purpose.

NORTH DOWNS COPPER MINE

In Redruth ; is a very old mine, and was working from 1718 to 1758, and made upwards of £100,000 profit.   It has been resumed of late, and making large returns, having sold in twelve months, ending June 30th, 1842, one thousand seven hundred and seventy-one tons of ore for £12,379 12s.   Arseniate of lead has been found in this mine.

HALLENBEAGLE COPPER MINE.

In Kenwyn, near Chacewater, is returning large quantities of ore, but barely paying the expences of working : about two hundred persons are employed—the monthly cost averages £1,500—and the machinery on the mine is valued at £4,500. From June 1834 to June 1842, the mine returned seventeen thousand one hundred and forty-eight tons of ore yielding £77,604 2s. 6d.

WEST WHEAL JEWEL TIN AND COPPER MINES

In the parish of Gwennap, were formerly known as Tolcarne and Roselobby. The rich lodes of Wheal Jewel run through the centre of Tolcarne, and the sett is in the vicinity of the Great Consolidated Mines and Wheal Unity. The present proprietors have been at work about six years, and have sunk to a depth of 80 fathoms, cutting several lodes as they went down ; and from 1837 to June, 1842, have returned ores to the amount of £12,305. 13s.,—the cost to the proprietors during that time having been upwards of £40,000. At present, they are rather more than paying the cost of working (£600 a month), with prospects of soon doing much better.

TRELEIGH CONSOLIDATED COPPER MINES

In the parish of Redruth, consist of Wheal Maria, North and South Good Success, Wheal Shanger, Wheal Christoe, and Wheal Fortune, and are in the centre of a very good district. About forty years ago, these mines were worked, and ore raised from above the adit, which has been driven through them, more than 40 fathoms deep.—There are several lodes in the sett, principally continuations of those in the surrounding mines.—The present company have been at work six years, having a lease for 21 years, from 1833, at 1-16th dues, and have sunk about 70 fathoms. At present, they are about paying the cost of working (£600. a month) and from 1836 to 1842, returned ores to the amount of £26,872. 18s. 5d. The cost to the proprietors, during that time, having been about £25,000.

BULLER AND BEAUCHAMP MINES,

In Gwennap, at a former working made a profit of £80,000, on an original outlay of £640. ; but, at present, they are worked on a very limited scale, and at a loss, being continued, in the hope of a discovery as they are situated in a good mining district. In the four years ending June, 1842, they sold 5,715 tons of ore for £32,398. 9s.

TEHIDY COPPER MINE

In the parishes of Redruth and Illogan, is about 50 fathoms deep, and turning out small quantities of ore ; having, in four years ending June, 1842, sold 926 tons for £5,973. 5s. 6d.

c

THE MINING DISTRICT OF CAMBORNE AND ILLOGAN.

Is comprised within a tract, bounded on the east by the valley, which divides Illogan from Redruth, on the south by the ridges of Carn Brea Cairn Entral, and Camborne Beacon Hill, on the west by a line from Camborne Beacon Hill, to about half-a-mile north of Camborne Church, and on the north by a line parallel to the highway from Camborne to Redruth.

The rocks of this district consist of an elevated range of granitic hills on the south; covered on their northern slope by varieties of slate and intersected by elvan courses, and by numerous lodes and cross courses. The alternations and mixtures of granite and slate are very remarkable, and this district affords many excellent examples of elvan courses.

The principal lodes of this district have a bearing varying from 20 deg. to 40 deg. south of west, some dip north, and others south; the main lode, and Harriott's lode in Dolcoath, the principal lode in North Roskear, two of the chief lodes in Cook's kitchen, and the principal in Tincroft, dip towards the south; but in the same mines, and in East Wheal Crofty, there are other lodes inclining to the north, which have also been very productive.

There are several Caunter lodes (or lodes bearing from 10 to 30 deg. north of west) in this district, and the greatest produce of East Crofty, Wheal Crofty, and Dolcoath, have been derived from them; but the principal operations are on two parallel lodes, one passing through Stray Park, Dolcoath, Cook's Kitchen, Tincroft, and Carn Brea, and the other through North and South Roskear, Wheal Crofty, East Wheal Crofty and East Pool, yellow copper is the largest product of this neighbourhood, though several mines have yielded native copper, and also grey and black copper ore; tin ore is also raised in large quantities from Tincroft, Carn Brea, Cook's Kitchen, and Dolcoath, and small quantities from East Pool.

In one of the lodes at Dolcoath, native silver, and vitreous and silver ore were found, and a piece of plate was manufactured from it, and presented to the late Lord de Dunstanville by the adventurers in that mine, and is still preserved in the family as an heir-loom. Cobalt and bismuth have also been found in Dolcoath, and in the lodes of North Roskear and East Wheal Crofty, small quantities of mineral pitch have been discovered

CARN BREA TIN AND COPPER MINES.

Consisting of Wheal Fanny, Tregajoran, Druid, and Burncoose, are in the parishes of Redruth and Illogan at the bottom of Carn Brea Hill, and are as rich and productive as any in Cornwall. Wheal Fanny many years ago, in conjunction with Tincroft, left a profit of several hundred thousand pounds.

Tregajoran and Fanny, a few years ago, were worked by William Reynolds, Esq., of Trevenson and party, but abandoned with a loss of £40,000, on account of the "wringing up" or contracting of the lodes. Capt. Joseph Vivian and Capt. Joseph Lyle then took the sett, and commenced driving towards the Druid, and having continued the workings through the contracted lodes; discovered several rich courses of ore: and the present proprietors, since July, 1834, have made a clear profit of £107,500; and are now making more than £1,000 a month, over the cost of working, which is £4,000. There are eight or ten lodes in the sett; eleven steam engines at work, which, with the other machinery on the mines, are valued at £60,000; and the mines give employment to 1,000 persons.

The deepest shaft is 140 fathoms, but the principal workings are carried on above the 110 fathoms level. In twelvemonths, ending in June, 1842, they sold 10,104 tons of copper ore for £67,734 16s. 2d. and tin for £4,000, which, at a cost of £48,000, would yield a profit of £13,734 16s. 2d.

EAST POOL COPPER AND TIN MINE

In Illogan, has been at work about ten years, and upon an original outlay, by the proprietors (128 in number) of £640, has returned ores, to the amount of near £130,000, out of which, besides the cost of working, (£1,100 a month), several thousand pounds have been expended in the purchase of machinery now on the mine, and a profit of £30,000 divided. The mine is 90 fathoms deep, 300 persons employed, and the present profits about £4,000 a-year. In twelvemonths ending June, 1842, were sold 2,943 tons of ore, yielding £21,172, being over the average price of Cornwall.

DOLCOATH COPPER MINE,

In Camborne, is one of the oldest in Cornwall, having been worked with but little interruption for near a century; it is 300 fathoms deep, and has left a profit of £600,000.

In Pryce's *Mineralogia Cornubiensis*, published more than fifty years ago, Dolcoath is mentioned as one of the most extensive and most important mines in Cornwall; the depth then was barely 100 fathoms; and the opinion of miners, at that time, limited the productiveness of copper lodes to a depth varying from 40 to 80 fathoms; and, although the ore was known to exist at a greater depth, it was considered to be deteriorated in its quality, and scarcely worth pursuing; but that this was a fallacy, many of the best mines sufficiently testify.   In Wheal Abraham, for instance, tbe lode at the 240 fathoms level, was larger than it was nearer the surface.   In Nov. 1814, a large cavern was discovered in Dolcoath at the depth of 170 fathoms from the surface ; its form was very irregular, from 18 to 20 fathoms in length ;   three fathoms high, and from four to nine feet wide ; in the lower part, and wedged between the walls, there are several rocks, between which are spaces which communicate with other cavities below.   In the "Valley" the workings are carried to such an extent, that no timber can reach from side to side in the levels, and still the lode is found to extend to a greater width.   The miners work in a swing stage, which they drop against such parts of the side as they intend to take away ; and then letting themselves down by means of a swing chain ladder, they blast down immense quantities of rock.   In 1810, silver ores were raised in this mine to the value of £2,000.

In four years, ending June, 1842, the mine returned 14,829 tons of ore yielding £68,638 16s., which left a small profit.   About 200 persons are employed.

SOUTH WHEAL BASSETT MINE

In Illogan, has been at work ten years ; the outlay by the proprietors (64 in number) has been £2,624. and from June 1834 to October 1842, the mine has returned copper ore to the amount of £106,331., and divided a profit of £13,760.   About 300 persons are employed, and the monthly cost £1,300.   The present profits are at the rate of £5,760. a-year.

NORTH ROSKEAR COPPER MINE

In Camborne, was first opened about fifty years ago, and left a profit of £90,000.   The present proprietors commenced working in 1819, and have made a profit of near £100,000., the present profits, however, are very small.   The mine is two hundred fathoms deep, about seven hundred persons employed ; the monthly cost £2,200. and the machinery on the mine valued at £15,000.   From June 1834 to June 1842, they returned ores to the value of £213,320. 1s. 6d.

SOUTH ROSKEAR COPPER MINE

In Camborne, on the same lodes and adjoining North Roskear, is working to a small profit, and employing four hundred persons.   From June 1834 to June 1842, this mine has returned ores to the value of £105,771. 19s. 6d.

WHEAL SETON COPPER MINE

In Camborne, is in the immediate neighbourhood of Carn Brea, Tincroft, East Pool, Cook's Kitchen. and Dolcoath Mines.   The lodes of East Wheal Crofty and North Roskear running into the sett. The great counter lode in Wheal Seton averages six to eight feet wide, and runs through East Pool, East Crofty, and three hundred fathoms of North Roskear; there are several other lodes in this sett. The proprietors, now ninety-nine in number, commenced working in 1834, have expended on the mine £18,339., and returned ores to the amount of about £900. The monthly cost averages £150.

EAST WHEAL CROFTY MINE.

IN Camborne, on East Pool, West Crofty, and North Roskear lodes, has made large profits; and now leaving a profit of about £300. a month, at a cost of £2,400. The principal workings are in green stone; near one thousand persons are employed; and in four years ending June 1842, the mine returned copper ores to the value of £130,859. 19s. 6d.

COOKS' KITCHEN COPPER MINE

IN Camborne, is a very old mine two hundred fathoms deep, and has left a profit of £300,000. Some years ago, the miners had driven several levels home to a cross head, which they took for the cross course, and not caring to open the ground too near the Tincroft sett, which was then full of water, they suspended their operations, leaving this cross head untouched; of late years, however, they have extended their workings in that direction, having a very productive lode. About two hundred and fifty persons are employed on the mine; and in three years ending June 1841, they sold by public ticketing, three thousand eight hundred and ninety-six tons of ore for £16,481. 15s., besides ores consigned to the miners company.

TINCROFT TIN AND COPPER MINES

IN Redruth. This and the adjoining mine, Wheal Fanny, now belonging to the Carn Brea mines, have produced at different workings upwards of £1,300,000. The sett is very extensive, and in the neighbourhood of Carn Brea, East Pool, and East Crofty, and on the estate of Penhillack, formerly the residence of Mr. John Hichens, where the mine encroached so much into his garden, that he was at length driven entirely away. The dues paid from one acre of ground were enormous, and they had the richest and largest course of tin under it, that was ever seen. The mine recommenced working in November 1833, and from 1834, to June 1841, has returned eight thousand nine hundred and forty-six tons of copper ore (besides tin) for £34,434 14s. 6d. The outlay by the present proprietors has been £42,000; they have divided a profit of £3,000., and the mine now leaving a profit of £500. a month. Cost of working £1,200. a month.

WHEAL ST. ANDREW COPPER MINE

IN Gwithin, near Camborne, was commenced from the surface, in 1836, and has returned one thousand six hundred and seventy-three tons of ore, yielding £7,923, 13s. 6d. down to June, 1842. It is seventy fathoms deep and leaving a small profit, the monthly cost being about £450.

SOUTH WHEAL FRANCIS COPPER MINE

IN Illogan, was worked many years ago, and again commenced in 1834. The length of the sett is eighty fathoms less than a mile on the course of the lodes, and more than a mile north and south.   The mine is bounded on the east by South Wheal Bassett, the lodes of which are running into the sett.   When the working commenced in 1834, the adit was driven on the South lode, nearly two hundred fathoms. Within the last month or two, an engine has been erected and set to work, and the cost to the adventurers (one hundred and twenty-four in number) up to June 1842, was £2,232.   When sinking twenty fathoms from the surface, several large pieces of oak timber were found in a fine state of preservation, after having lain there upwards of a century.

UNITED HILLS COPPER MINE

IN St. Agnes, forty years ago was worked as Wheal Rock, but the lode was very hard, and a *rock* back at surface, instead of gossan, and the mine was abandoned in 1816 with a heavy loss.   In 1826 it was re-commenced, and in 1836, made a scrip company, with an estimated paid up capital of £20,000.   From June, 1836, to 30th June, 1842, the mine has returned ores to the amount of £118,110. 8s. 6d. and paid a profit of £18,500.; the first dividend having been paid in Oct. 1836, and the last in October, 1841.   At present the mine is leaving a small profit.   There are several lodes in the sett, the run of one (*a counter,*) passes through Wheal Fanny, and is visible to the eye at surface. They have an eighty inch cylinder steam engine, &c. at work, and employ about four hundred persons.

WHEAL CURTIS TIN AND COPPER MINE

FORMERLY worked as Wheal Drim, is in the parish of Crowan, to the south of Crenver and Wheal Abram, which when worked left a profit of £150,000. to the proprietors.   In 1836, operations were commenced by Capt. Teague, and carried on solely by himself till his death in 1840, he having expended upon the mine, £15,000. and returned ores to the amount of £5,110. 15s. 6d.   In August, 1841, the workings were again resumed, and in twelve months they returned ores to the amount of £2,693. 11s. The sett is one mile square, with five lo les in it, although they are only working upon one; the engine shaft is forty fathoms below the adit of twenty fathoms, and they have an engine of forty-five inch cylinder, and a water wheel for stamping and dressing the ores.

TRENOWETH COPPER MINE,

In Crowan, is on the same range of lodes as Wheal Abraham, Crenver, and Oatfield, which have left, from time to time, profits of upwards of £150,000. This mine was, some years ago, worked by a party who expended some £20,000, in extending the adit-level, nearly to the centre of the sett, longitudinally, and in sinking several shafts in the main line of the lodes; one of which, the engine-shaft, is 40 fathoms under the adit, or about 80 fathoms from the surface, as is also the whim-shaft; from the former, a cross-cut, north, will intersect the lodes, if extended 15 or 20 fathoms, and a south cross-cut from the same point, it is computed, would expose the south lode, which is spoken highly of by the old working miners. The operations about the adit-level were attended by expositions of good ore ground; the returns from 1814 to 1822 being 4,976 tons, yielding £19,520.; but, for reasons which are not easy of explanation, the mine was abandoned in the latter year, although we are informed that the establishment might, even at that period, have realized a return to meet the current expenditure; and several hundred pounds worth of ore have been raised since that period, by tributers. There is a large counter lode running through the sett, which intersects the other lodes near the centre of the grant; and here, also, is the junction of the granitic and argillaceous schistus formation, which is generally looked upon by miners in that neighbourhood with great interest. And this, together with the general character of the lodes, has induced a very highly respectable party to resume the workings under the mining management of Mr. Oliver H. Matthews, whose good government of various extensive mining establishments abroad, ensures them, if economy and skill can realize the desired object, a successful issue to the undertaking.

THE TIN MINES OF ST. JUST AND ST. IVES.

THE mines and minerals of St. Just (chiefly tin) are comprised within
a district of the north-western coast, of not more than three miles long,
and one mile and a half in breadth ; but, after giving a slight sketch of
this district, other tin and miscellaneous mines will be given under this
head, viz., some of Breage and Marazion.

St. Just is on the western shore of Cornwall ; its cliff scenery is of
the boldest character. Its mines, being almost all either in the slate
(which, where it occurs, is seaward), or in the granite, many of the
levels are extended beneath the bed of the ocean, the adits usually
opening so far above the sea-level, as to be out of reach of the waves.
The labourers descend to the entrance either by a zig-zag path, as at
Levant, or by ladders in the face of the cliff, as formerly, at Botallack.

Of its minerals, bismuth has been found in Botallack and Levant,—
silver in Levant—iron ores, in great variety, in Botallack, Wheal
Edward, Wheal Cock, &c.—cobalt in Botallack—pitch-blende and
granite in Wheal Edward—oxinite, hornblende, apitite, epidote, and
jaspery iron ore in Wheal Cock—calc-spar and arragonite in Levant—
schorl in Botallack—and opal in Wheal Maitland. But the directions
of the various veins are the most remarkable features of this district ;
the metalliferous veins bearing from north-west and south-east, to north
and south ; whilst the cross-veins run about north-east and south-east,
or not very different from the direction of the lodes in other parts of
the country. There are but few instances of a lode, which has been
cut off by a cross course, being honestly recognized on the opposite
side ; so that the " heave" of one vein by another, which, in other
parts of Cornwall, occasions no uneasiness for the chance of re-
discovery, is here of considerable moment and uncertainty ; and the
hardness of the rock (country) is an obstacle to the shafts being sunk
perpendicularly, and to the extension of levels at right angles to the
lodes (cross-cuts), for the discovery of parallel branches.

In Little Bounds and Wheal Cock, the excavations have been made
into the sea ; in the first, it being on the beach, and dry at low water,
it was secured with well-caulked planks ; in the second, the hole made
by a borer, being but small, it was stopped by a plug. The site of
some of the steam-engines on the top of the cliffs, is highly picturesque.
The water from most of the mines is used for domestic purposes, and
sometimes fifty women may be seen at once standing around an
engine-house, washing the linen of their families in the warm water
from the steam-engine. This particularly occurs at Boscean and
Boscaswell Downs, and Wheal Owls ; and it is rather a singular sight,
of a Monday-morning, to see the females hastening to the mines,

bearing on their heads their washing-trays, and the linen of their respective families.

In this district there are several stream works for tin, the principal of which is *Carnon*, situated near a navigable branch of the river Fal, which receives several rivulets from the hills of Stythians and Gwennap. The vale of Carnon is 300 yards in breadth, the hills on the south being rather steep, whilst, on the north, their declivity is very gradual. The largest quantity of tin ore was found where the Stythians Vale, and a glen called Smelter's-House Vale, open into the Carnon Vale. Small pieces of gold have been found among the tin ore in this stream. Tin to the amount of £8,000. a-year was formerly raised from this vale,

The Drift Moor stream work is near Penzance, forming part of a valley which opens at Newlyn. This valley was explored for tin, at a very remote period; an ancient water-engine having been found by the last tinners, buried in the alluvial soil.

BOTALLACK TIN AND COPPER MINE,

In St. Just, is upwards of 150 fathoms deep from the surface; and at a former working, left a profit of £300,000. There are several lodes in the sett, now returning tin, and large quantities of very rich copper, averaging £14. per ton—making a profit of £1,000. a-month. The workings are upon the cliff near the Land's End, overlooking the Atlantic, and are carried out 480 feet beyond low-water-mark; and, in some places, so near the working of the sea, that the heaving of the waves is distinctly heard by the miners. The mine was wrought under the sea, beyond the memory of any person now living; and worked so high as to open a communication between the water and the mine, which is now stopped by a wooden platform, on which is laid a mass of slimy turf, and the whole covered by the stony fragments of the beach. The situation of its steam-engine, near the base of the cliff, and within a few fathoms of the sea, was formerly one of the sights of the west. This mine was formerly worked solely for tin; and, not paying the expences, was upon the point of being abandoned in 1841, when they discovered rich courses of copper, which have since returned 1,269 tons, yielding £15,906. 13s. 6d.

BALLESWIDDEN TIN MINE.

In St. Just, in Penwith, about three miles west of Penzance. The sett extends over a large tract of land, containing five principal parallel tin lodes, besides several counter lodes, and branches of tin leading from one lode to another. The mine commenced working in 1832; is

90 fathoms deep from the surface ; and, from 1833 to 1841, returned tin to the amount of £51,960. 8s. 6d., and at present selling about 50 tons per month.   There are five steam-engines, two water-wheels, and six horse-whims at work, and the machinery on the mine is valued at £18,572.   The dues only 1-33d, and the lease for 21 years, from 1841.

LEVANT TIN AND COPPER MINE,

INCLUDING the old mines of Zawnbrinney and Boscriggan, in St. Just, have left a profit of more than £150,000 in the last eight years.   The copper ore of this mine is the richest in Cornwall, and it also returns small quantities of tin.   They have an engine of 40-inch cylinder, and four steam-whims ; the adit is 25 fathoms deep, and the lode is more than 220 fathoms deeper than the adit ; the most productive part of the mine being under the sea.   The steam-engine is situated on the top of the cliff, and the spectator standing at the adit's mouth, near the sea level, almost shudders as he looks up at the engine-house, and other appendages, seemingly suspended over his head in the air.

The roaring of the sea in stormy weather, and the breaking of the waves on the beach, is distinctly heard by the miners.   The water in the mine is salt, but there is very little of it.

A new lease for 21 years was granted to the present proprietors (80 in number) in 1839 ; and the mine is making small profits. About 600 persons are employed ; and the monthly cost, £1,800.   In eight years, ending June, 1842, this mine has returned 21,374 tons of copper ores, yielding £255,538., besides large quantities of tin.

BOSCASWELL DOWNS TIN MINE,

IN St Just, is an old mine, and an intervals has given considerable profits ;   a single bunch of tin, once gave the adventurers from £15,000. to £20,000.   The mine is about 200 fathoms deep, and worked by a small engine, but very little doing.

BOSCEAN TIN MINE,

IN St. Just, is worked about 60 feet deep, in two shafts ; an engine of 24-inch cylinder, drawing out of both of them.   The adit is 14 fathoms ; little doing.

ST. IVES CONSOLS TIN MINE,

IN St. Ives, has been at work 27 years, and is very rich for tin, 170 fathoms deep, and has made large profits.   Monthly cost is £2,400, and employ 450 persons ; but, owing to the low price of tin, the mine is barely paying expenses.   The formation of tin ore in the *St. Ives*

Consols, which is provincially called the Carbona, joins the standard lode at the depth of about 80 fathoms, and the part by which it is united is not more than a few inches square.   From that place it has been worked, perhaps 120 fathoms in a south-easterly direction, until, tending continually downwards, it reaches the depth of nearly 100 fathoms.   Its greatest height is about 10 fathoms, and its largest breadth about the same; but the average dimensions may be 4 fathoms high by 10 or 12 feet wide.   Its bulk, however, is subject to very great irregularity.   It exhibits few of the usual characters of a *lode,* as it is bounded above, below, and on either side, by the usual granite ; and it has an irregular dip of from 45 deg. to 80 deg. towards the south-west.   It is chiefly composed of felspar, quartz, schorl, and tin ore, but in many places it contains fluor chlorite, common and blistered copper pyrites, iron pyrites, and vitreous copper ore.   About 80 fathoms from the *lode,* it falls in with, and takes the direction of, the *middle trawn,* and they continue side by side (the *carbona* on the west) for about 25 fathoms.   The *carbona* then becomes a little mixed with the substance of the *trawn,* and in a few fathoms farther, the very felspathic disintegrated granite of which the latter is composed, is gradually entirely replaced by a mixture of schorl, quartz, and tin ore, closely resembling the composition of the *carbona,* as long as the direction continues that of the *trawn.*   At length the *carbona* takes a bend, and goes off at right angles towards the north-east ; whilst the *trawn* continues its course, and gradually becomes more granitic, although it still retains traces of the *carbona* for some fathoms ; but at last it reassumes the felspathic and disintegrated characters common to the *trawn.*

Immediately east of the *trawn,* the *carbona* increases to enormous dimensions, and is worked for at least 10 fathoms in length, breadth, and height.   The scattered lights, the great number of miners in their soiled and torn working-dress, the pillars and beams of wood which support the roof and walls, the rock lining this vast cavern, all dimly and at intervals discerned by flickering and uncertain gleams, produce a very striking and uncommon effect.   From this large mass, *shoots, branches, pipes, bunches,* and other irregular protuberances, strike off in every direction ; and wherever there is an unusual enlargement of the *carbona,* it is invariably found to occur at the crossing or union of two or more of these, which may be called the *limbs.*   Some of them bear about east and west, and two of them, called *Williams' and Kempe's lodes,* have rather more resemblance to veins than is usual in the *carbona ;* but these are for the most part cut off by the granite above, below, and at either end, and these abrupt terminations are often not on joints in the rock.   Some of these *bunches* pass off diagonally from the main body, and, when followed, lead to other strings, which are extremely irregular in their direction, and are also cut out by the granite. These again send off *shoots*, which also often open into short vein–like masses, and these may in their turn throw off strings both to the *carbona*, and to its subordinate branches.

Throughout there is a regular mineral transition from the substance of the *carbona* to the contiguous granite ; and the composition thereof, like that of the *lode*, seems equally to depend on the nature of the *country*. For where the granite is cross-grained, with large crystals of white felspar, there is but little tin ore ; and this, on the contrary, abounds where the granite is more uniform in its aggregation, and the felspar crystals pink or pale green, and not very clearly defined.

The whole may be described as a net-work of *pipes, strings, branches, shoots, and veins*, converging into one grand trunk, which extends to the south-east, and dips, in the same direction, about 1 in 6 ; on all sides surrounded by the hard granite, but nowhere extending to the surface.

WHEAL VOR TIN MINE,

In the parish of Breage, 3 miles from Helston, has been the richest tin mine in Cornwall ; the first steam-engine erected in the county, was on this mine, between 1710 and 1714. There are now fifteen engines at work on this extensive sett, which has the appearance of a town : the machinery is valued at £100,000 ; and more than £200,000 profit have been divided among the shareholders. The lode from which the chief part of the ore has been raised, is still productive. A few years ago, the monthly cost of working was £12,000, but it is now very much reduced : they employ 1,200 persons, and make a small profit.

There is a blacksmith's forge in the bottom of this mine, 1,470 feet below the surface of the earth, now in full operation, at which all the miners tools are steeled, sharpened, and repaired, bucket-rods cut and welded, and many other necessary jobs done. The smithy is clear from dust, smoke, and sulphur, and does not in the least annoy the miners at their work ; there are also smelting works on the mine, where they smelt their own tin, and return about 150 tons a-month.

GODOLPHIN TIN AND COPPER MINE,

In Breage, was formerly under the management of Messrs. Williams, of Scorrier, and left a profit of £50,000. It was then idle for twenty-five years, till 1836, when it was purchased by the late Capt. Teague, and worked till 1841 ;—the cost expended upon it was £40,000, beside the returns, which amounted in copper alone, to £49,131. 10s. 6d. After Capt. Teague's death, the mine was purchased by Capt. Lyle and

Mr. Grout, and they have lately cut a very rich lode. In this mine the lodes (both tin and copper) run nearly north and south.

WHEAL BUDNICK TIN MINE,

In Perranzebuloe, is a very bunchy mine, sometimes rich, and, at others, poor. It is bounded on the west by Wheal Leisure and Great St. George, on the north by North Wheal Rose, and on the south by Perran Consols. The mine is between 50 and 60 fathoms deep ; the adit, 34 fathoms : and first made returns in 1834, and paid the first dividend in the latter part of that year, and has divided a profit up to the present time, of £12,000. The present monthly returns are about 20 tons, which, owing to the low price of tin, does not pay the cost of working. About 200 persons are employed.

PERRAN AND GREAT ST. GEORGE MINES,

In Perranzebuloe, more than fifty years ago, a grant of the Perran sett, was given to the Great St. George adventurers ; but they, having abandoned it, the grant became void,—and in 1830, Capt. John Vivian obtained the sett for ten years, and soon made a discovery of tin and copper ; which, becoming known, the Great St. George Company applied for an injunction, and for a short time stopped Capt. Vivian from working. The injunction, however, was soon dissolved, and a law suit commenced between the two companies, and continued for ten years, ending in favor of the Perran, just as the lease of the sett had expired. During the ten years, upwards of £70,000 worth of ore was raised from Perran, which would have left large profits, had it not been for the law suit. The Great St. George and Perran are about to be resumed in conjunction with Bolenna, Wheal Leisure, and Droskin Mines ; at a former working, Bolenna left a profit of £90,000 in eight years, and Wheal Leisure left £60,000 in about the same time.

POLBEROW TIN MINE,

In St. Agnes, was formerly one of the richest tin mines in Cornwall, but, at present, is not paying the cost of working. In 1750, the tin ore in this mine, consisting principally of large grained crystals, was so abundant in several parallel and contiguous veins, that they could not find horses enough in the neighbourhood to carry it to the smelting-house, and were compelled to take it away in carts. A considerable part of these ores were so rich that it did not require to be stamped, and large blocks of tin-stone were obtained. In March, 1750, one was taken to Killenick smelting-house, weighing 664 lbs., and was so rich that it yielded 11½ for 20, without stamping or dressing. Another block brought to the smelting-house, from the same mine, weighed 1,200 lbs.

PRINCESS ROYAL TIN AND COPPER MINE,

In St. Agnes, close to the Great George and Wheal Leisure Mines; the sett contains several lodes, and more than £10,000. have been expended in sinking shafts, draining levels, &c , and a new company have lately commenced working.

WHEAL VIRGIN COPPER MINE,

In St. Hilary, is 120 fathoms deep; the principal workings at present being in the " bottoms." The 110 fathom level has been driven about 130 fathoms, and the main dependence of the mine is now upon the south lode. In 12 months, ending December, 1841, they raised 2,045 tons of ore, yielding £12,651. 9s. 8d., and, after paying Lord's dues, £533. 18s. 2d., and the cost for working, £12,357. 9s. 3d., the mine was in debt for the 12 months' working, £239. 15s. 9d.

WHEAL PENROSE LEAD MINE,

In Sithney, has lately been purchased of the British Silver Lead Mining Company, by a party of gentlemen, and set to work. The machinery and materials are worth £2,000, The British Silver Lead Company expended upwards of £100,000. on this, and Wheal Unity Mine, and returned a large quantity of lead; and, in 1841-42, 225 tons of copper. yielding £1,350.

TREWAVAS COPPER MINE,

In Breage, on the edge of the cliff, two miles west of Porthleaven, in a former working left a profit of £1,920. There are four or five lodes in the sett, lying close together; the deepest level is 70 fathoms. From 1835 to Oct., 1842, the mine has returned 10,838 tons of ore, yielding £71,981. 13s., and divided a profit of £7,680. About 200 persons are employed, and making a profit of £4,000. a-year.

LESSEAVE COPPER MINE,

In Breage, near Trewavas, has lately been set to work, and not sufficiently developed to justify an opinion respecting it. It has returned 36 tons of copper ore, yielding £287. 2s.

WHEAL DARLINGTON TIN AND COPPER MINE,

In Ludgvan, was formerly worked as the Bog Mine, and is about 110 fathoms deep; working in two lodes, and leaving a profit of about £8,000. a-year; the accounts are held every three months. The materials in the mine are valued at £12,000., and the prospects considered to be good. Three hundred persons are employed; and in three years ending June, 1841, returned 7,435 tons of copper ore, for £48,308. 10s.

TREVARNO TIN AND COPPER MINE.

In Sithney, was worked fifty years ago, but not to any profit; a steam engine has lately been erected, and the mine again set to work, having returned ores, in 1841, to the amount of £349. 14s., at a cost of £2,150. 3s. 8d.

TREGOTHNAN CONSOLS MINE,

Near Falmouth, is not paying the cost of working, the ores being very poor; they have returned 995 tons of copper ore, yielding £3,496. 6s. 6d.

THE MINES OF MARAZION.

The principal mines in this district were worked a few years since by Mr. Thomas Saunders Cave. They consisted of Great Wheal Fortune, Rospeath, Wheal Bolton, Owen Vean, Wheal Prosper, Wheal Friendship, &c. &c.—all being on branches of the same great Champion lodes, and Gwallon on parallel lodes to the south. The total loss by Mr. Cave on twenty-seven mines amounted to £192,722. 14s.

Great Wheal Fortune, formerly called after an utensil of great domestic use; was worked by the great grandfather of the present Sir Charles Lemon, Bart., and made a profit of £100,000. During the working of Mr. Cave, it yielded ores to the amount of £50,000, but, owing to the extravagant outlay, was abandoned with a loss of £29,658 11s. 11d.

Rospeath has been worked for copper to a depth of fifty-seven fathoms; the lode is a continuation of that in Great Wheal Fortune. This mine yielded ore, but was abandoned with a loss of £15,637 18s. 10d.

Wheal Bolton is separated from Rospeath by a very large cross course, and the mine has been worked for copper to about eighty fathoms under the adits of seven fathoms. There are two lodes, which dipping different ways, are heaved the same way by a cross course; the loss on this mine was £15,637 18s.

Gwallon is forty fathoms under the adit of fifteen fathoms, and is a very old tin mine. Mr. Cave's loss on this mine was £16,485.

Owen Vean was worked in 1750, and left large profits. Mr. Cave's loss upon it was £22,234. 3s. 4d.

Wheal Prosper was abandoned with a loss of £23,804 13s. 10d.

The Marazion Mines, consisting of Wheal Virgin, Wheal Maid, Rodney, Crab, and Tregurtha Downs, at an old working, left a profit of £40,000; and about seventeen years ago were resumed, and from 1831 to 1841 returned ores to the amount of £150,000, and were then abandoned with a loss. Wheal Crab, at the western part of these mines, where a valuable discovery was made and an engine erected, has lately been resumed under the name of WHEAL CHIPPINDALE, in compliment to Charles Chippindale, Esq., of St. John s Wood Road, Regent's Park, and is in a fair way of making a profitable mine ; both tin and copper ores occur, but the latter by far the most abundantly, and down to October, 1842, has returned one hundred and fourteen tons of copper ores, yielding £700. 7s.

The only patch of granite within this district, which is one of the most interesting in Cornwall, is St. Michael's Mount. That part of its base which lies towards Marazion is of slate, whilst its summit and southern side are of granite. The junction of the two rocks is well shown in the beach, as are also the granite veins by which the slate is traversed ; some of these are almost entirely micaceous, containing topazes, apalite, crystals of tin ore, and other rare and curious minerals, and some of these may be traced in both rocks. There are numerous fine elvan courses in this district, which traverse the slate ; the first occurs in Great Wheal Fortune Mine, and in and about it, much of the produce of that mine occurred ; a second in Gwallon, and a third in Marazion mines. One of these is again seen in Wheal Darlington, and at the back of Penzance Pier, along the beach, between that place and Newlyn, and.through the *Wherry mine*, where tin ore of £70,000. value, was raised from it.

Many years ago, St. Michael's Mount, now surrounded by the sea, was distant five or six miles from it, enclosed with a very thick wood, and called in Cornish " *Caraclowse in cowse*,"—*the hoare rock in the wood*, and it is supposed, that by a convulsion of nature in 1099, the submersion of the wood and adjoining land took place. Whether or not this may in any degree account for many of the submarine mines now existing, is a matter not yet determined. Among the most extraordinary of the latter is the Wherry Mine.

THE WHERRY MINE,

NEAR Mount's Bay, Penzance, as a submarine mine, is celebrated as the most extraordinary in the history of Cornwall ; and was established on a shoal 720 feet from the beach at high water ; the rock is covered about ten months in the twelve, and the depth of water on it at spring tides 19 feet ; and in winter the sea bursts over the rock in such a manner as to render all attempts to carry on mining operations unavailing ; veins of tin were first discovered in an elvan course running in front of Penzance in the early part of the last century, and attempts were made to work it, but abandoned as hopeless. Notwithstanding all the difficulties, however, a poor miner of Breage, named Thomas Curtis, in 1778, had the boldness to renew the attempt, and after innumerable difficulties succeeded in forming a water-tight case, as an upper part of the shaft, against which the sea broke, while a communication with the shore was established by means of a wooden frame bridge.

As the work could be prosecuted only during the short period of time when the rock appeared above water, three summers were consumed in sinking the pump shaft; and the use of machinery becoming practicable, the water tight case was carried up a sufficient height above the highest spring tides. To support this boarded turret, which was 20 feet high above the rock, and two feet one inch square, against the violence of the surge, eight stout bars of iron were applied in an inclined direction to the sides, four of them below, and four of an extraordinary length and thickness above. A platform of boards was then lashed round the top of the turret, supported by four poles, which were firmly connected with these rods. Upon this platform was fixed a wins for four men. But it was found impossible to prevent the water forcing its way through the shaft during the winter months; or remove the tin-stone from the rock to the beach; the whole winter, therefore, was a period of inaction, and it was not till April, that work could be resumed. In the autumn of 1791, the depth of the pump, shaft, and of the workings was 26 feet; the breadth 18 feet: 12 men were employed for two hours at the wins in hauling the water, while six men were teaming from the bottoms into the pump. The men worked on the rock six hours afterwards—in all eight hours; and thirty sacks of tin-stone were taken in an average every tide; and ten men in the space of six months, working about one-tenth of that time, broke about £600 worth. After a time a steam engine was erected on the green opposite, and hanging rods from it, carried along the wooden bridge to the mine, and in this manner, ore to the amount of £70,000 was raised from it. And the treasures were not exhausted at its close, which was as romantic as its commencement. An American vessel broke from its anchorage in Gwavas Lake, and striking against the stage, demolished the machinery, and thus put an end to an adventure, which, both for ingenuity and success, was probably never equalled in any country,

In 1836, a Company was formed, and the mine again set to work; but not paying the Company was broken up in 1840.

THE EASTERN DISTRICT.

A few years ago, the idea that ore in any quantity could be found in this district was laughed at by the miners of the west, who seemed to have a fancy to confine mining to the neighbourhood of Gwennap ; but a few enterprising individuals have set the question at rest by opening some of the richest mines in the county.

In this district will be included the mines of St. Austle, Callington, Liskeard, and Tavistock. The St. Austle district is principally stanniferous, the copper lodes being chiefly confined to its south eastern side, as in Fowey Consols and Crinnis ; the neighbourhood of Callington, is both stanniferous and cupriferous, with a few silver lodes. At Wheal Duchy, near Callington, a silver lode was found at the 10 fathom level, worth £200 per fathom, and from which £3,000 worth was raised. Dartmoor is stanniferous ; the mines around Tavistock are cupriferous. On the north are the lead lodes of Wheal Betsy and Lidford, and on the south the argentiferous lead mines of Beer-Alston. In 1812, silver was raised from Wheal Brothers to the amount of £3,000. On the north of Tavistock there are manganese mines ; and manganese has also been obtained near Calstock. Iron as a brown hœmatite, is found of rich and excellent quality near Lostwithiel and Bodmin ; and cross courses holding iron ore occur in a north and south direction at Nantallan Downs, and Lanivet Downs near Bodmin ; and between 1778 and 1802, 9293 tons of iron were shipped from a lode near Combe Martin to the iron works at Llanelly in South Wales.

The lodes in this district generally run east and west ; but a powerful cross course traverses Wheal Franco, heaving the lodes to the right, taking a direction thence towards the south western part of Tavistock.

Several fine lodes of rich hœmatite iron ore are found in this district, which have been worked at various times, and micaceous iron ore is found at various places upon Dartmoor, and under the name of *Devonshire sand*, was some years since used in London as writing sand, selling at £3 3s. to £8 8s. per ton. Magnetic iron ore, of good quality, has been worked near South Brent, Devon, and at Penryn, Cornwall, And, according to Sir H. T. de la Beche, a large supply of good hœmatite iron ore could be readily obtained from this district, if the necessary demand existed.

A great many new mines have been set to work, particularly in the neighbourhood of South Caradon ; but the majority of them are not sufficiently developed to report upon, only the principal, therefore, will be noticed. In the neighbourhood of St. Austle, there are several stream tin works, the principal being *Happy Union* in Pentuan vale, near St. Austle. This valley is a continuation of St. Austle Moor, where for ages a great quantity of tin has been got by streaming. In some parts it is 600 feet in breadth, in others not more than 300 ; the tin ground lies on the solid rock, and is generally from three to six, and sometimes 10 feet in thickness ; the quantity of tin ground opened at Pentuan had been 18,200 square fathoms, and the average of block tin got per square fathom has been 186 pounds ; the quantity of over-burthen removed has been upwards of 200,000 tons. Happy Union was first opened in 1780 ; Wheal Virgin there forming a part of it. The former being worked towards the sea, and the latter towards St. Austle Moor. Near here, there are also,—the *Merry Meeting* stream works,—*Rock*, near St. Austle—the *Grove—Watergate*, &c., &c.

FOWEY CONSOLIDATED COPPER MINES

IN the parish of Tywardreath. In 1813, these mines, then called Wheal Treasure, Wheal Fortune, and Wheal Chance, commenced working ; and stopped in 1819. The amount expended on them, during that time having been £49,563. 16s. 11d. In 1822, they were purchased by J. T. Treffry, Esq., of Fowey, and consolidated under the above title. In 1836, Lanescot, an adjoining mine, was united with the Fowey Consols, this mine having divided £45,000 between 1822 and 1832.

From August 1815, to the end of 1841. the Fowey Consols Mines returned 234,486 tons, 8 cwt., 2 qrs. of copper ; fetching £1,422,633 17s. 1d., out of which the profit paid to the adventurers, with a reserve fund not divided, amounted to £179,995 11s. 6d. The value of the stock on the mines, engines, materials, &c., is £60,000. There are six steam engines at work, of altogether 302 horse power ; seventeen water wheels, of 484 horse power ; and three hydraulic engines of 119 horse power. There are five pumps, or engine shafts in course of sinking, the deepest being about 200 fathoms below the adit of 45 fathoms. There are 20 lodes now in course of working, the principal of which run through the sett for nearly two miles. 1,792 persons are employed upon the mines ; and the system of working is principally for discovery ; large quantities of ground being constantly opened, and large piles of ore left as a reserve, while operations are carried on in search of more.

PAR CONSOLS MINE.

NEAR Fowey Consols, also principally belongs to Mr. Treffry ; and is one of the richest mines in Cornwall. The mine first returned ore in the latter part of 1840, and from that time to Oct., 1842, has yielded 9197 tons of copper ore, which sold for £68,624 1s. and is now making a profit of about £10,000 a-year. The machinery on the mine is worth £5,000.

CHARLESTOWN UNITED MINES.

IN St. Austle, 3½ miles from Fowey Consols, are very rich for both copper and tin, and, in 1841, returned ores to the amount of £44,000. At present they are making a profit of between 4 and £5,000. a-year. The cost of working is about £2,500. a-month ; they are now sinking a shaft in the copper mine, and expect shortly to cut the lode. The upper part of the great tin lode in this mine, is open to daylight for several fathoms, on one side ; and, from the same lode, great quantities of copper were formerly taken in Crinnis Mine.

OLD CRINNIS COPPER MINE,

IN St. Blazey, near St. Austle, was several times abandoned before it became profitable, and, in the year 1808, was declared, by one of the best Cornish miners, Captain James Michell, to be " not worth a pipe of tobacco !" In 1809, Mr. Joshua Rowe, of Torpoint, and co-adventurers, commenced working again, notwithstanding the general unfavourable opinion of the mine ; and, as she still remained poor, one by one the adventurers dropped off, leaving the entire cost of working upon Mr. Rowe, who, after laying out a few hundreds in a fresh part of the sett, discovered a rich mass of ore at about ten fathoms from the surface. Upon this becoming known, all the old adventurers (who, in the poverty of the mine, had thrown up their shares, and refused to pay cost) again claimed their shares—which claims, however, Mr. Rowe resisted, and a law suit, lasting for some years, was the consequence. Mr. Rowe ultimately obtained a verdict in his favour, and, in the short space of four years and a half, the mine made a clear profit of £168,000., besides paying £20,000. law expenses. After 1815, the mine gradually became poor, and was at last abandoned ; and has not since been worked.   Alderman Sir Matthew Wood, Bart., M.P., was one of the fortunate co-adventurers with Mr. Rowe, and Mr. B. Wood, M.P. for Southwark, for many years had the management of the mine.

POLGOOTH TIN MINE,

NEAR St. Austle, is a very old mine, and for several years made a profit of £20,000. a-year.   It has been worked some time under the management of Mr. Taylor, and for some months made profits, but at present not paying expenses.

SOUTH CARADON COPPER MINE

IN the parish of St. Cleer, near Liskeard, was originally searched for tin,[1] and when the lode was first discovered in Caradon Hill, and found to contain a quantity of gossan, it was considered so unfavourable to the existence of tin, that it was with difficulty a Company was formed to work it ; but the Messrs. Clymo who had obtained the sett, persevered, and three rich copper lodes were soon opened. The original outlay to the adventurers before the mine made returns in August, 1837, was only £327 8s. 5d., and from that time to the 31st March, 1840, they sold copper ores to the amount of £15,635 10s. 7d., paid all costs for machinery, including two steam engines and a whim ; and divided a profit of £2,400, leaving £1,379 8s. 10d. in hand : from that time to November, 1842, they have divided, altogether, a profit of £19,168, and are now receiving at the rate of £10,000 a-year, with every prospect of greatly increasing the returns. Some mine agents have asserted that there is £150,000 worth of ore discovered in this mine ; but be that as it may, the prospects are exceedingly brilliant, and not surpassed by any other mine in Cornwall. A great part of the workings are in Caradon Hill, which is 1,208 feet high. The monthly cost of working is about £1,800.

WEST CARADON COPPER MINE

In St. Cleer, and near South Caradon. The extent of this sett is 400 fathoms from north to south ; and 100 fathoms from east to west ; it extends further north and south than the South Caradon sett, and will take all their lodes. The mine has lately been set to work, and has already returned 375 tons of ore for £2,737 1s.

CARADON CONSOLS COPPER MINE

In St. Cleer, and near South Caradon. There are old workings in this mine, but no one can remember them being worked. The present Company have lately commenced, and hope to equal South Caradon, being upon the run of all her lodes.

WHEAL ROBINS COPPER MINE

Near Liskeard, three miles west of Caradon Consols ; they have an adit driven 70 fathoms on the lode, and have sufficient water to work her 100 fathoms. Not yet made any returns.

TOKENBURY COPPER MINE,

In Linkinhorne, nearly adjoins South Caradon, to the south, and has lately been set to work, the cost to the proprietors, 128 in number, has been £2,560. The run of the lodes is from South Caradon, east to Tokenbury ; in some parts of the mine the granite is overlaid by killas. The adit levels were commenced in killas, and are now in granite ; more than a mile of backs of lodes, have been laid open, and a little copper discovered near the surface.

MARKE VALLEY TIN AND COPPER MINES,

In Linkinhorne, six miles from Liskeard. The sett extends over 120 acres, and is held under four grants for terms of years. The engine shaft is only 25 fathoms below the adit.

CORNWALL GREAT UNITED MINES,

In Linkinhorne, near South Caradon. £70,000 to £80,000 have been expended on these mines ; and they are just making returns from a fine course of ore in Clanacoomb.

TRETOIL TIN AND COPPER MINE

NEAR Bodmin, has been worked four years, is between 40 and 50 fathoms deep, and upon an original outlay of £1,000, has returned from September, 1838, to the 6th October, 1842, 6306 tons, 6 cwt., 2 qrs., of copper ore for £40,743 19s., and 7 tons of tin for £257 1s. 1d., and has paid the adventurers a profit of £3,000. The value of the machinery on the mine is £1,000 ; and it is returning about 130 tons of copper a month, besides a small quantity of tin, at a cost of £900.

An adjoining sett, formerly called Blackheath, and then East Tretoil, has been lately added to the Tretoil, and in future they will be worked as the Tretoil Consolidated Mines.

TREGOLLAN COPPER MINE,

In Lanivet, two miles from Bodmin, was formerly called Wheal Change ; the present Company commenced working in 1836, and have expended upwards of £21,000 : the returns have been 614 tons of ore, yielding £2,722 ; but owing to the discovery of a fine course of ore in the shaft, the returns are expected to increase greatly ; and the proprietary has been consolidated into 135 shares.

BODWANNICK COPPER MINE,

NEAR Bodmin, has lately been set to work. The length of the sett is 610 fathoms ; they have three copper lodes discovered, and are driving on the course of them, into a hill, the same as they did at South Caradon.

HOLMBUSH COPPER MINE,

IN the parish of Stoke Climsland, near Callington, was, some years ago, worked to a depth of 70 fathoms. The present proprietors commenced working in December, 1834. The sett is held under the Duchy of Cornwall, and a new lease for 21 years was granted in Dec. 1840. The mine is managed by a Board of Directors in London, and the profits divided every two months. The cost to the present proprietors has been £14,000, and they have already made a profit of £20,000. The deepest level is 110 fathoms from the surface; and it is calculated they have at least £50,000 worth of ore in sight; the ores are rich, averaging from 9 to £10 per ton, and about 210 tons a month are raised at a cost of £1,000, dividing a profit of £6 000 a year among the proprietors. There are three steam engines on the mine, the machinery is valued at £6,000, and about 250 persons employed. Their first sale of ore was on the 21st June, 1836, and from that time to Oct. 1842, they have sold to the amount of £101,243 4s. 5d.

WHEAL FRIENDSHIP TIN AND COPPER MINE,

In Mary Tavy, near Tavistock, is in clay slate, and worked to a depth of 200 fathoms entirely by water power; for this and the adjoining mine, Wheal Betsy, a fall of water of 526 feet in height is employed in giving motion to seventeen overshot wheels; eight of them performing the duties of pumping water from a depth of 200 fathoms; the diameter of the largest of these wheels being 51 feet, with a width of breast of 10 feet clear within the rings; the smallest of the eight being 32 feet in diameter. Four other wheels give motion to machines for drawing up the ores to the surface, their diameters varying from 40 to 26 feet; and the five remaining wheels are employed for mills for crushing and stamping the ores. In addition to all this power, a steam engine, with a cylinder of 80 inches diameter, and 10 feet stroke, is provided as an auxiliary in periods of drought or frost. The principal workings are on the 150 fathoms level east of the engine shaft—160 fathoms level west; the 128 at Brenton's shaft, and the 70 and 80 driving east.

They sold in 12 months ending 31st Dec. 1841, tin and copper to the amount of £31,836. 14s. 5d, leaving a profit after deducting every expense of £4,893 2s. 6d. About 150 persons are employed.

WHEAL BETSEY LEAD MINE,

Adjoining Wheal Friendship, is 110 fathoms deep, worked by water power, and returning near 45 tons of lead per month. She is barely paying expenses. The amount of ore sold in 12 months ending 31st Dec., 1841, was 663 tons, 2 qrs. for £7,207 3s. 9d., whereas the cost for working amounted to £7,196 9s. 2d,

This mine was re-opened in 1806, and about the year, 1821, was returning 400 tons of lead, and from £4,000 to £5,000 ounces of silver annually.

COOMBE MARTIN MINES.

In North Devon, were worked in the reign of Edward the 1st., and Edward 2nd., and made great returns of silver; they were again opened in the reign of Elizabeth. In the time of the Commonwealth they were recommended to the notice of Government, and were again worked at the end of the 16th century, but without success. In 1813, they were re-opened, and, in four years, produced only 208 tons of lead ore. In 1837, they were again set to work, and are at present making good returns, having divided £5,000 profit in two years. The mines are 67 fathoms deep.

DARTMOOR CONSOLS TIN MINES

CONSIST of Wheal Gobbett, Deby Hole, and Wheal Cumpston; the two former mines are in the forest of Dartmoor, in the parish of Lidford, Devon. The sett is more than a mile in length from east to west, and in some places more than half-a-mile in breadth Wheal Gobbett was openly worked for stream tin, by the ancient tinners, for 300 fathoms in length. The present proprietors have driven an adit-level on the course of the lode, about 150 fathoms, and have returned about £1,000. worth of tin. They have a water wheel of 24-feet diameter, worked by a water-course from the river Swincombe; and the mine is about 40 fathoms deep. Wheal Cumpston is in the parish of Holne, and was formerly extensively worked for tin.

WHEAL FRANCO COPPER MINE,

NEAR Tavistock, commenced working in 1823, and has returned about £60,000. worth of ore; but, at present, not paying expenses. The sett is held on lease for 21 years (16 unexpired) at 1-12th dues. The old engine-shaft is about 120 fathoms deep; and lately, in sinking a new shaft, they cut a rich course of ore. The mine is worked by water wheels placed on the bank of the river Tavy, three miles from Tavistock, on the Plymouth road.

GUNNIS LAKE MINE,

IN Calstock, on the banks of the river Tamar, is celebrated for its rare and beautiful varieties of copper ore, specimens of which are to be found in most of the mineral collections in the kingdom. It has been worked by several companies, and has given large profits, having yielded more than £250,000. worth of ore. There are several lodes in the sett, which have not been worked under the adit, and it is confidently expected that these lodes will be productive of wealth to Mr. Noah Coward and his friends, who are now working them. The old mine is 100 fathoms deep from the adit of 55 fathoms, but not worked to any extent. The lodes, going east, pass through Tamar, Wheal Luscombe, and Wheal Bedford, the Canal and Crown Dale mines, which have also given large profits. The lodes, going west, are supposed to pass through Holmbush and Caradon mines ; the ores in the latter strongly resembling the Gunnis Lake. The present party have been working the mine with steam-power, but, in future, they hope to bring in a sufficient water-power for every purpose. The country about this mine is singularly beautiful ; and in the opposite hill, which joins Gunnis Lake, will be found

THE BEDFORD MINES,

UNDER the management of Mr. Josiah Hitchens, and are worked by water power, a large wheel is erected, and working the Marquis lode ; another wheel is in progress of erection, which will enable the adventurers to work the north part of the mine to a considerable depth. There must have been large returns of ore from this mine, as the lodes are entirely worked away for 200 fathoms above the deep adit. The present proprietors have expended £8,000.

DRAKE WALLS TIN MINE,

ADJOINS Gunnis Lake to the south, and is a very ancient and productive tin mine : hitherto it has been worked by water, but that being now insufficient, a 40-inch cylinder steam-engine is now in the mine, ready for erection.

TAMAR SILVER LEAD MINES,

ON the banks of the river Tamar, in the parish of Beer Alston, Devon, are worked to a depth of 145 fathoms. The outlay to the proprietors has been £18,000., and they have divided a profit of £4,500. They have three steam-engines, and several water wheels at work. The lead is very rich for silver, yielding on an average 60 to 65 ounces of silver per ton of lead ; and, in the year 1841, returned ore to the amount of £14,646. 12s. 2d., at a cost of £12,067. 6s., leaving a profit of £3,579. 6s. 2d. The monthly cost averages about £850.

THE MINES OF CUMBERLAND, &c. &c.

In Cumberland, Westmoreland, Yorkshire, and Derby, the principal mines worked, are for lead, and carried on chiefly by local adventurers. In Cumberland, and other places, there are many remains of old workings for copper and other minerals, in some of which, ore of a richer quality than in Cornwall is found, and many from their great size, richness, and facility of operation, might be worked to great advantage; and it is wrong to suppose that only the lead mines of Cumberland are worth working. A rich vein of copper has lately been discovered at Haygill, on Caldbeck Fells, where, in driving for the main vein, a rich string of ore, from seven to eight inches in depth, was found. There is also a copper mine near Hesket Newmarket, 14 miles south of Carlisle, called Currock-end, which, 30 years ago, produced some thousand pounds worth of copper ore. The sett is very extensive, and several lodes in it, but never has been worked to a profit. The engine shaft is only 20 fathoms deep from the surface, and working with a 30-feet water wheel.

Edward IV., in the 8th year of his reign, granted all his copper mines, containing gold and silver, in Cumberland, Westmoreland, and Northumberland, to Dodrick Waverswick; and Dr. Fuller, in his " Worthies of Cumberland," observes, " that in taking the rich copper mines from the Duke of Northumberland, at Keswick, it came to pass, that this Queen (Elizabeth) left more brass than she found iron ordnance in the kingdom." And in the 10th year of her reign, Plowden says, " She took from the Earl of Northumberland, his rich copper mines of Keswick, because of its holding so much silver and gold in the ore."

Boyle, in his " Useful Philosophy," says, " A friend of mine found, in his own land, a parcel of ore, which seemed to be copper. After fusion it yielded very good copper, but the person to whom he committed the examination, being extraordinary skilful, found, besides the copper, a considerable quantity of silver, and in that silver a good portion of gold."

The grants made by Edward III., of the copper mines in Northumberland, Cumberland, and Yorkshire; those of the same King, of *all the gold and silver mines* in Cumberland, Westmoreland, and Devonshire; those of Richard II., Henry IV., and Henry VI., of the same mines; a grant of Henry VI. to John Boatright, of all copper, tin, and lead, possessing *gold* and *silver;* more especially the grant of Henry VII., who, in the first year of his reign, constituted the Duke of Bedford, and others, Commissioners of *all* the mines, not only of gold and silver, but of tin, lead, and copper, in England and Wales, upon paying to the King the fifteenth part of the gold and silver, and to the lord of the soil the eleventh part; and also the grant of Queen Elizabeth, who, in the 6th year of her reign, granted all ores, mixed and compound, and all other metals, minerals, or treasures to be found in earth or ground in England, and the English pale in Ireland, with licence to dig and search for the same, and build houses, &c. &c., for their own use, (the disturbers of the grantees or their miners, to be imprisoned for six months, without bail or mainprize) with power to take up and use all necessaries for their workings, and to *fell timber*, &c. &c.; " the Queen to receive, for every hundredweight of gold ore, eight ounces of gold, and of the silver, the twentieth part;"— were all, as Mr. Abbott says, very prejudicial to the progress of mining, and operated directly contrary to the intention and expectations of the Crown and grantees, who had the power, not only, of seizing upon people's mines and working them, but of destroying their land and estates; for though there were clauses in several of the Crown grants, obliging the lessees to repair any damage done to the estates, the grantees seldom or never paid any attention to the proprietors of the land, but tore it up, sunk shafts, and made roads at their pleasure, and left the proprietors to their remedy—so that, instead of leading to the discovery of new veins, or the increase of the number of mines and workings upon those veins already known, these grants caused a general inactivity and stagnation, and mining in these counties became neglected.

In all the old grants of Queen Elizabeth to the incorporated society of " Mines Royal," and also the society of " Mineral and Battery Works," which were confirmed by James I. to the Earl of Pembroke, Lord Cecil, and others, the counties of Cumberland, Westmoreland, Yorkshire, Lancashire, Devon, and Cornwall, and the principality of Wales, were granted, as being " the counties where gold and silver mines were most abundant."

An author of some experience in mining operations, in 1700, expresses his surprise " that more pains are not taken to search *Essex* for mines of different metals,"—and several learned people assert that the money which Cunobeline, Prince of the Trinobantes, coined at Camelodunum, in Essex, was drawn from a mine in that county. This is certainly a mistake, although some have appeared to believe it; and Agricola, in his work, " De Re Metallica," p. 26, speaking of those parts, says, " Naturalia venarum signa observavi," but here he did not allude to the usual metalliferous deposits, but veins of Ruddle and iron pyrites, some of the latter of which, it is possible, might occasionally produce very small quantities of gold.

Henry IV , in the 22nd year of his reign, having received information of a concealed gold mine in Essex, commanded Walter Fitzwalter to " apprehend all persons concerned in concealing the said mine, and bring them before the King and his Council, to receive what shall be ordered."—*Tower Records, Rot.* 34.

About 160 years since, two gold mines were stated to have been discovered; one at Pollux Hill, in Bedfordshire, and the other at Little Taunton, in Gloucestershire, " The Society of Mines Royal *seized them*, and granted two leases of them to some refiners, who extracted some gold; but they did not go on with the work, as the gold sometimes would not repay or requite the charge of separation, *though often it did.*"—*Essay on Metallic Works.*

Dr. Leigh, in his " History of Lancashire," page 82, says, " We have, in England, quantities of copper sufficient to supply all Europe. If the mines of copper ore were rightly managed, we should not import *any* copper." Also, page 82, " The right method of running the *copper, which is got in the north,* is by reducing the ore to a small powder, afterwards by wasting it, and then, by an addition of lixivial ashes, the ore, in a proper furnace, will run into a *fourth* part of malleable copper."

The Beacon Rake lead vein, in Derbyshire, was wrought by the Romans with great advantage, and, several centuries after, it was partly re-opened by an ancestor of the Duke of Rutland, who made very large profits from the workings, but was compelled to abandon them in consequence of the then-prevailing ignorance of the geology or habitudes of the strata about the mine.

Consumblock, in Cardiganshire, is stated to have been wrought by the Romans and Saxons, and, afterwards, with great profit, by the Patentees of Royal Mines; and after them, by " The Company of Mine Adventurers." The vein is now larger than ever, and now yields above 60 ounces of silver in a ton of lead.—*Thiers' " Second Discourse."*

The Greek and Roman miners were held in high estimation, and the ancient civil law granted them many privileges; alleging, as a reason for so doing, that the miners were very useful and valuable to the public.

Queen Elizabeth, in the tenth year of her reign, by letters-patent, discharged all miners, and all other persons occupied in finding, digging, and refining metals or minerals, from paying any taxes or impositions, and relieved them from serving on juries, and made them free from arrests.

There are numerous lead veins in the West Riding of Yorkshire, and the principal part of the mines are in the range of hills between Pately Bridge and Buckden; those lying near Pately, or towards the south-east border of the mining district, are the Corkhill Mines, held by lease of Sir Thomas White, and the Providence, Prosperous and Merrifield Mines,—the three last all on the same vein, and worked by one Company. These mines, with several smaller ones, carried on by poor miners, have for some years produced about 1,200 tons of pig-lead per annum.

A few miles north-west of the above, are the Grassington Mines, on his Grace the Duke of Devonshire's property, and worked on his own account.

The ores from the Pately district are smelted in or near the mines, in ore hearths, or open blast furnaces; and those from the Grassington district, in reverberatory furnaces. None of the ores contain sufficient silver to pay for refining.

The PARYS MINE, in Anglesea, was almost a mountain of copper, and the very streams were impregnated with metal to an extent which made it profitable to precipitate it from them; but the mine was worked out, and no such a discovery has since been made in the neighbourhood

In the neighbourhood of High Peak, Derbyshire, a capital of £100,000. is invested in mining at the present time.

The MOLD MINES, in Flintshire, returned in the twelve months, ending Dec. 1841, 1,413 tons 9 cwt. of lead, yielding £17 763. 9s. 2d., which, after deducting costs of working, and lord's dues, left a profit of £3,086. 7s. 4d. There are several mines in this sett, the principal one being Pant-y-mwyn, which alone yielded a profit, in the past twelve months, of £3,184. 0s. 5d.,—the loss from the other mines reducing the general profit, as before stated. This mine returns about 100 tons of lead per month, aud is about 100 fathoms deep.

The GOGINAN MINE, near Aberystwith, produces from 180 to 190 tons of lead ore per month, and is not much more than 20 fathoms deep. In the year ending Dec. 1841, the quantity of lead ore raised was 1,528 tons 19 cwt., yielding £22,068. 14s., — leaving a profit of £988. 17s., after deducting cost for working, and Lord's dues. The dividends paid in 1841 amounted to £5,000.

The LISBURNE MINES, in Cardiganshire, about 40 fathoms deep, returned 1,153 tons 2 cwt. of lead ore in 1841, yielding £12,221. 12s. 5d. which did not pay the cost of working.

The GOGERDDAN MINE, in Cardiganshire, returned 232 tons 14 cwt. of lead ore, in 1841, yielding £3,373. 19s. 8d., while the cost of working for the same period was £6,968. 6s. 7d.

According to the statistical return of Mr. Taylor, in the year 1835, the lead mines of Cumberland, Northumberland, and Durham, returned 16,626 tons of ore—Yorkshire, 4,700 tons—Derbyshire, 4,000 tons—Shropshire, 3,539 tons—Devon and Cornwall, 140 tons—Flintshire

9,380 tons—Denbighshire, 177 tons—Cardiganshire, 1,200 tons.   Since
which, the returns have greatly increased.

At the Eagle Crag mine, at Adderdale, Cumberland, in 1837, a vein
of lead ore was discovered, one knocking of which weighed upwards of
30 tons.

There is a very rich copper mine at Eardiston, near Oswestry,
Shropshire.   The rock in which it is worked, being a redstone, the ore
is principally the green carbonate, mixed with earthy black ore and red
oxide.   Some of this ore has yielded a produce of 14¾ per cent. of
copper.   About £8,000. have been expended on the mine, and an
engine, machinery, &c., erected, worth £2,000.   The sett is a mile in
length on the course of the lodes, but the operations are confined to
about 75 fathoms in the centre, between two cross courses.   The deepest
level is 45 fathoms.

A tunnel has lately been completed through Llandudno Mountains,
in North Wales, leading to some copper works ; under the superin-
tendenee of Mr. Thomas Jones and Captain Davey.   The tunnel was
commenced in February, 1834, and has been worked by 12 miners
alternately, night and day.   It is cut in a straight line, measuring
874 yards, and is supplied with air by a wind-pipe, through a shaft,
cut from above.   It is constructed with an arched roof, and is
6½ feet high.   On approaching the boundary of the old works, the
miners drilled a hole through the side, which found its way into the
confined water, in about 5½ feet.   Two additional holes were then cut
above, and instantly the water burst through with great velocity, in
a stream of not less than 396 gallons per minute, and ran through the
tunnel like a cataract.   It was computed that a body of water, 198 feet
in depth, and of immense breadth, was vacated by this means from the
works.   When completed, it is anticipated the tunnel will lead to a
bulk of copper ore, not yet explored.   The proprietors are, E. Lloyd,
Esq., of Cefn, and W. and A. Worthington, Esqrs., of Whitchurch.

It would appear from the preceding, that, with the ancients, mining
was principally carried on in the northern counties, but the Cornish
have outstripped them in the spirit of enterprize ; and to Cornwall,
alone, the attention of speculators and capitalists is now directed :
although, if " bonnie " Cumberland would but shake off her lethargy,
and *prove* that her mines are worthy their attention, she is not so " far
north,'' but that a slice of their patronage might reach her miners.

THE HISTORY OF METALS.

In giving an account of the metals, it is not the compiler's intention to divide them into their classes and orders, or into any methodical arrangement; but to give just such a general summary of them, as may interest the miner, without leading him into the intricacies or mysteries of science.

To prove the great antiquity of metals, we are informed by Holy Writ, that the children of Adam were both acquainted with the existence, the utility, and the art of working them; the early mode of reducing the mineral, or purification of the crude or native metal, being simply (as it is supposed) conducted with wood burned in square ovens, made with potter's clay. Josephus observes, with reference to the uses of clay, that when Seth and his children, had been forewarned of an universal deluge by Adam; they erected pillars of stone, and brick, upon which they engraved the various inventions and discoveries that had been made in the arts and sciences, since the creation of the world, and Pliny and Strabo are in error in ascribing the first knowledge of the art of working clay either to Greece or Egypt. The manufacture of metals into various devices, applicable either to husbandry, warfare, or as objects of devotion, likewise claim a contemporaneous antiquity; Tubal Cain, one of the sons of Lamech, being mentioned as the first artificer of copper (or brass, as it was then called) and iron. The Phœnicians, however, were the first who conducted mining with any system; they extended their commerce over the greater part of the known world, and established several cities in countries where their mining operations employed a sufficient amount of population to require a fixed settlement. From several accounts we learn, also, that they most probably discovered the wealth of this island, and carried on an extensive commerce in tin and copper with the inhabitants. The first classification of metals which we have recorded, is, that by Avicenna, an Arabian philosopher, who wrote the " Commentaries on Aristotle," in A.D. 1030. He divided them into four classes —stones, metals, sulphur, and alkalies.

We find no other trial made till the year 1546, when Agricola completed a work, entitled, " C. Agricola de re Metallica, et de Natura Fossilium;" this, however, contained but little new information. The next work on the subject was the first arrangement of Linnæus.

But to rerurn to more modern times, the valuable statistical account of the copper trade in the year 1691, with other matter connected with the metal trade, has been obligingly communicated for this work, by Mr. Thomas Irving Hill, late of Redruth.

TIN

Is a metal of the greatest antiquity, in fact, the oldest of the metals, first found in Spain and England, and known by the name of metal of Jupiter. Tin has never been found pure, but only in a state of oxide and sulphuret (the latter being very rare) and it often contains small portions of iron and silex. In one vein, in Cornwall, an ore has been found, called bell-metal ore, from its resemblance to that metal in colour, and consists of tin, copper, and sulphur, and a small portion of iron. Tin is found principally in those rocks, which from their not containing any animal or vegetable remains are called primitive ; it occurs in beds, but principally in veins, accompanied by the ores of arsenic, iron, copper, and zinc, also quartz, mica, fluate of lime, and other substances. It is also found in Cornwall in rounded portions, in alluvial beds, or depositions from the ruins of rocks.

It is found most abundantly in Cornwall, but also in Saxony, in Bohemia, Malacca in Asia, in Chili in South America, in China and in France.

It is supposed the Cornish first became tinners upon the arrival of the Phœnicians on their coast, at least 500 years before the Christian era, for though tin was discovered before that time, its value seems to have been unknown. All is conjecture as it regards its first discovery ; but, as Cornwall was formerly very thickly wooded, the inhabitants having no coal, used brushwood and turf as their only articles of fuel, and for the most part they cut their turf from the Moors, where they did not confine themselves to the surface, but went to a considerable depth after *peat*, and, as the Cornish Moors once contained an abundance of tin, it is supposed, that in this way it was first discovered, and being washed from the different lodes, by successive floods, valuable deposits were accumulated in the different brooks and rivers, and other cavities of these moors ; and to the present day, what is called stream tin is held in the highest estimation by smelters.

Until 1838 all the tin of Cornwall paid a duty of 4s. per 120 lbs. to the Duke of Cornwall, or to the Sovereign when there was no Duke ; and the fees to the officers and loss of time were equal to 1s. more. The duty is now abolished, and the miners reap the benefit.

Until the latter part of the 17th century, all the tin produced in Cornwall, was smelted in blast furnaces with charcoal. It was not until the former part of that century that pit-coal had been successfully ap-plied to the smelting of any of the metals. The decrease of wood in Cornwall, and the consequent increasing expense of smelting with charcoal, naturally induced the tinners to turn to any substitute, and to try the use of pit-coal. This was the cause of the erection of air reverratory furnaces, in which the fuel and the ores were separated, and culm coal mixed as a flux with the ore.

The first air furnace for smelting tin was erected about 1680, and since that period nearly all the mine tin of Cornwall has been smelted in air furnaces.

There are now three kinds of tin made in Cornwall, viz., grain tin, refined tin, and common tin.

Grain tin was formerly made solely in blast furnaces, only from the diluvial tin ores, or what is generally called stream tin, remarkable for its superior purity.

It was the only kind of tin used for making tin plates (or, rather, for tinning the plates of iron), on account, partly, of its fluidity, and partly of its superior colour and lustre. It was also used, in small quantities, in dyeing scarlet and in making tinfoil.

Grain tin has lately been almost wholly made in reverbaratory furnaces, like other tin, but still generally from diluvial ores. The cheaper mode of manufacture has greatly reduced the price.

Refined tin, though not equal in quality to grain tin, is made from selected ores, and fluid enough for the first coat on iron plates, and is used by most of the tin-plate manufacturers.

Common tin is made from the mass of tin ores of Cornwall.

The name, *grain tin*, is given from its quality of granulating. This is done by placing a block of it in a furnace kettle, and heating it as high as it will bear, without melting; it is then raised by a pulley to a considerable height, and suddenly dropped on a hard surface, by which it becomes instantly divided into small striated masses, to which the name, *grains*, has been given.

The name, *block-tin*, has long been disused.

The quantity of grain tin consumed in dyeing, and in making tinfoil, is nearly, if not quite, 200 tons per annum.

Putty is the oxide of tin elutriated,

The tin mines of Cornwall have not, on the whole, been a profitable concern. The year 1837 was a peculiarly unprofitable one. The state of the tin mines in that year was ascertained for the information of the Government, on an application from the miners for the abolition of the duty paid to the Duke of Cornwall. The result was as follows :—

Loss on fifty-eight mines £111,517
Deduct supposed increase in the value of property in the Mines... 31,000

£80,517
Profit on ten mines................ 20,358

Net loss........... £60,159

F

When the tin mines just pay the expenses of working them, about one-half of the cost of the tin may be reckoned as paid to the workmen, and the other half for agency, expense of draining the mine, coal, timber, machinery, &c.

From the year 1750 to 1837, the annual produce of the tin mines in Cornwall never exceeded 5,000 tons, but generally averaged from 2,500 to 3,500 tons ; and the average price paid to the tinner, in forty years, from 1746 to 1788, was 64s. 4d. per cwt.

In 1787, Mr. George Unwin, a purser of an East India ship,—as an adventure,—took some tin from the Malacca Isles to China, and made a handsome profit by the speculation. On his return to England, having learned the price of tin in Cornwall, he brought the subject before the East India Company ; and, in 1789, the East India Company purchased and sent out a small quantity from Cornwall, which fully answered their expectations ; upon which they entered into arrangements with the tinners of Cornwall for an annual supply. This exportation to India speedily advanced the price in Cornwall ; but the Cornish having found the benefit of such a connexion, were not easily induced to relinquish it. An artificial system was, therefore, created, by which the East India Company were still supplied, although their price was lower than that paid to the tinners in Cornwall, whilst the price in the home market was kept high enough to make up the deficiency. By this system, the quantity delivered to the East India Company had always reference to the produce of the mines, and the demand at home,—this, varying from 500 to 1,500 tons per annum, and the average price of tin in Europe was, in consequence, much higher than it would otherwise have been.

The first price paid by the East India Company was £68. 13s. 4d. per ton, delivered on board their ships in London. In 1792, it was advanced to £71., and, on the renewal of their charter, about that period they agreed to take as much as 800 tons annually, at £75., and a further quantity of 400 tons (should the Cornish wish to sell it), at £68. 13s. 4d. These prices continued until 1809, when the difference between £75., and the price of the home market was so great, that the Cornish refused to sell any more unless the price were advanced, and, in 1810, none was exported to India. In 1811, the Company advanced the price to £78., and, in 1812, to £80. The system, however, became more and more difficult to maintain, and, in 1817, this connection with the East India Company terminated, and only a few small parcels were afterwards exported to India.

COPPER

Was first discovered and worked on the Island of Cyprus, from which it derives its Grecian name, and was denominated, by alchymists, the metal of Venus. The ores of copper are very numerous, and found in almost every mineral district in the world ; in beds, or more commonly in veins, accompanied by other mineral substances, as ores of zinc, lead, and tin, with quartz, fluate of lime, and spar in abundance. Native, or pure copper, is not found in any great quantity.

The most common ores of the Cornish mines is of a yellow colour, called yellow copper ore, or copper pyrites, and consist of copper, iron, and a large proportion of sulphur. It also occurs as native copper and red oxide, both very rare ; blue and green carbonates, and black and grey copper.

In combination with tin, it forms bronze and bell-metal ; with zinc, it forms Prince Regent's-metal, &c. ; combined with acetic acid, it forms verdigris ; and, with sulphur, the sulphate of copper, or blue vitriol. Brass is made of two-thirds copper, aud one-third zinc.

In Cornwall, copper was first discovered in working tin mines, in 1691, and a charter granted to Sir Joseph Hearne, and others, merchants of London, for purifying and refining copper, and incorporated under the firm of " The Governor and Company of Copper Mines of England," and now called, the English Company. In 1694, a copper coinage of halfpence and farthings, from Swedish copper took place, for which copper, Government paid 1s. 6d. a pound, equal to £168. per ton. In 1717, money was first coined from English copper, to the extent of 700 tons, for which the Government paid 15¾d. a pound, or £147. per ton. In 1762, the first brass works in the kingdom were erected at Bristol, and are still carried on by the Messrs. Harford. For the first twenty years of the last century, copper and brass utensils for culinary purposes were imported from Hamburgh, and made at the manufactory of Luremberg, and other places in Germany. So late as 1745, 1746, and 1750, copper tea-kettles were imported in large quantities from Holland.

The price of copper, from 1720 to 1722, ranged from £100. to £130. per ton, for cake, while the maximum price of manufactured was 1s. 6d. per pound,—the minimum price, 1s. 2d., or from £160. to £130. 12s. 6d. per ton. In 1731, the India Company first exported copper to India, and, till 1751, in cake only ; they then exported manufactured, for which they paid £135. 6s. per ton. Their whole exports, from 1731 to 1751, did not exceed 205 tons yearly,—from 1751 to 1772, they increased their exports, sending out cake at £105., and manufactured at £124. per ton. The average export in these years was 1,721 tons annually. In 1773, copper mines were first discovered in Derby and Wales, and this fresh supply so reduced the price, that, in 1781, the India Company purchased the cake copper at £79. per ton. In 1781, commenced the great opposition between the Anglesea and

Cornish copper smelters ; the latter making a sacrifice of £25,000. in the India contract, and to indemnify themselves for such a loss, they threw down the standard to such a ruinous price, for three successive years, that the Cornish miners were driven, as a remedy, to the acting proprietors of the Anglesea mines, to effect such terms as would bring the metal of both counties to market at a fair stipulated price. They entered into an agreement for seven years, which, however, in its effects, was of no advantage to the Cornish miners, as the price of copper, to the India Company, from 1781 to 1791, averaged £79. a ton.

In 1792, a great change took place in the trade; the consumers of copper in Birmingham formed two smelting companies, wishing to render themselves independent of others, in such an important article in their manufactures ; and, in consequence of this competition, there was a considerable rise in the standard, which continued to operate favourably on the price, shewing, the more immediate and direct the intercourse of the miner and consumer of copper, the larger profits accrue to the former ; whilst the latter hold a more independent position, and are entirely free from the caprice and intrigues of intermediate parties.

The first public sale of copper ores in Cornwall, on record, was in 1729, when were sold 2,215 tons, 12 cwt., for the year ; in 1764, the annual produce was 16,437 tons, giving 1,869 tons, 4 cwt. 3 qrs. 15 lbs. of metal, at a produce of 11⅜ per cent. ; in 1773, the amount was 27,654 tons, giving 3,152 tons, 1 cwt. of metal ; in 1779, 51,273 tons, giving 4,923 tons, 5 cwt. 3 qrs. 27 lbs. of metal, at a produce of 9⅝, and standard at £121. 8s. ; in 1820, 91,473 tons, giving 7,508 tons, 3 qrs. 26 lbs. of metal, at a produce of 8¼, standard £113. 15s. ; in 1841, the annual amount of sales had increased to 154,972 tons, giving 11,202 tons, 3 cwt. 9 lbs. of metal at a produce of 7¼, standard £125. 5d.

GOLD

Is one of the oldest metals, and most ancient writers have left documents of the produce of the mines of their time. It is of a yellow colour, or sometimes greenish, or reddish white, according to the metals with which it is combined, and is the most ductile and malleable of all the metals ; four hundred square inches of gold leaf, only containing *one grain and a half* of gold. It is mostly found in the metallic state, but generally alloyed by other metals, as silver, and copper, &c. It occurs in veins or beds, and in the former is generally accompanied by quartz, felspar, tin, silver, lead, and other metals. A great quantity of gold is obtained in grains, and round masses in the ruins of rocks, and in streams ; in this state it has been found in Wicklow, in Ireland, where, a short time ago, a quartzose and ferruginous sand was discovered, containing many particles of gold, with pepitas, or solid pieces, one of which weighed *twenty-two ounces ;* and no less than

1,000 ounces of gold were collected. A single specimen, a few years ago, was found among tin in a stream work (Carnon Vale) in Cornwall, equal in weight to upwards of ten guineas ; but the greatest quantity has been found in a fine sand, from the Peruvian, Mexican, and Brazilian rivers ; and from some of the African. In Europe, the Danube, the Rhine, and the Rhone, and the streams of Hungary and Transylvania, afford small quantities.

In France, the gold coin consists of 9 parts gold, and 1 of copper, and the jewellers' alloy is in three higher proportions. In England, the standard gold consists of 11 dwts., with 1 dwt. of copper. The alloy of gold and silver is seldom used, as a very small quantity of the latter whitens the colour of the former; whereas copper increases its brilliancy, and gives it a degree of hardness. The purple of cassius, used in colouring porcelain, is formed of an alloy. of gold and tin, treated with acids.

The ores of gold are reduced in the large way, either by amalgamating with mercury, or by fusion with lime and vitrifiable matter—the first process being applicable to those ores containing native gold, and the second to auriferous pyrites. Amalgamation is thus performed :—The ore haviug been brought to the surface, the larger masses are broken up with sledge-hammers, by the workmen, who afterwards reduce it into pieces not larger than a walnut ; these are sorted according to their supposcd relative value, and subsequently sent to the mill, where they are reduced to the state of fine powder. When this has been effected, it is generally washed, to separate as much of the light stony matter as possible ; the residue left from the washing, is to be dried and mixed with a sufficient quantity of mercury to amalgamate the gold and silver contained ; to favour which, a gentle heat may be applied to the mass for two or three days, at the end of which the fluid amalgam is to be poured off and pressed in a skin of leather,—this will separate a considerable part of the mercury, which is again applied to the same purpose, and distillation of the residue left in the skin. The gold is now left in an impure state in the retort, and may be purified by cupellation or gravitation. In this simple way, the greater part of the gold which comes from South America is obtained.

SILVER

Was worked at a very remote period in Greece, Macedonia, in the Black Sea, on the borders of the Rhine, and in Spain ; and was called the metal of Diana, or of the moon; and the alchymists esteemed it one of the principal objects in their theory of the transmutation of metals. It occurs in the pure and native state, but is sometimes alloyed by a proportion of gold, and sometimes copper. It is found in filaments disseminated through rocks, but chiefly in veins, in Peru, Mexico, Saxony, Bohemia, Norway, Hungary, and England. The ores of silver are numerous, and found combined with almost every other metal; the common lead ore, galena, mostly contains silver, but not always sufficient to pay for extracting it. (For mode of extraction, see *Lead.*) —Silver is not very hard, is very white, very brilliant, and malleable, and is applied to various purposes. In coinage, the French current silver contains 1-10th copper, and 9-10ths silver. The English alloy is 11-12ths silver and 1-12th copper.

The greatest part of the produce of silver is employed by the silversmiths, and the rest coined. The town of Birmingham, alone, consumes upwards of £90,000. worth a-year, in the manufacture of plated goods. The mode of plating is as follows :—A thin plate of silver is applied to a bar of copper, with a little borax between them, and the two, bound together, are exposed to a red heat; the borax melts, and the silver adheres to the copper; the bar is passed through the rolling-press, and thus comes out *plated*. French plating is performed by applying leaves of silver successively to heated copper, and fixing them by burnishing.

The assay of silver ores is extremely easy, nothing more being necessary (as silver is not acted upon by the fixed alkalies), after previously roasting, to separate any sulphur or arsenic that may be contained, than to rednce it to fine powder, mixing it with three or four times its weight of caustic potassa or soda, in a crucible, and applying a sufficient heat in a portable furnace, the metal will be found at the bottom,—if not pure, it is either to be capelled with lead, or fused with repeated additions of nitre in a crucible.

The silver mines of Argueros were first discovered in 1825, by a muleteer, who was cutting wood on the mountain. He found by chance some rolled blocks of native silver, and, on the discovery, some miners went to the place, and collected more than 10,000 piastres value, from the rolled stones they gathered on the surface. From that time, until 1840, they have produced annually about 30,000 marcs of silver, or about £120,000. These mines are worked on two lodes, which run from south east to north-west, and dip almost vertically, with a slight inclination towards the south-west. The run of these lodes is very regular, but the breadth varies from two to three feet.

The mules employed at the amalgamating mines in Mexico, are opened after death, and from two to seven pounds of silver are often taken out of their stomachs.

In Cornwall, silver is generally associated with galena, and has been found in several places.

IRON.

IRON mines have been wrought in this country from a very early period.

Those of the Forest of Dean were known in 1066, but in consequence of the great consumption of timber they occasioned for smelting, they were restrained by Act of Parliament, in 1580 ; soon after which, Lord Edward Dudley invented the process of smelting iron with pit-coal, instead of charcoal.

Iron is found in all soils, and in almost every rock. The ores are numerous, and are found in beds, in veins, and disseminated through rocks : it occurs combined with manganese, carbonate of lime, silex, alumine, sulphur, and oxygen. With copper, the arsenic acid, and silex, it forms a beautiful mineral, crystallized in cubes of a green colour, which are often transparent—it is called the arseniate of iron.

The principal quantity of iron is raised in Wales, Lancashire, Staffordshire, Shropshire, Derbyshire, Yorkshire, &c. &c. The various kinds of iron may be classed under the five following heads :—Cast, or pig-iron ; wrought, or malleable iron, and steel. Plumbago, commonly called black-lead, is a natural compound of iron with a large proportion of carbon. The richest ore of iron is the hœmatite.

Steel is refined iron, combined artificially with carbon. A bar of iron covered with charcoal, exposed to a strong heat, increases in weight, and comes out steel. Steel made red-hot, and suddenly cooled, acquires such extraordinary hardness as to cut iron, and even softer steel itself. Files sometimes are hardened to such a degree as to break by a fall.

The surface of a piece of iron can be made steel, and stopped before penetrating through the whole ; a paste of horn, or the charcoal of animal substances, covers the iron, and is wrapped in clay; in less than two hours, the iron comes out superficially steeled. The steel of gun-locks is thus made. The fiery sparks produced by the collision of flint and steel, are fragments torn off by the flint. By the violence of the stroke, they are heated red-hot; they brighten as they recede, and are brightest when at some distance. They are blown up into inflammation when passing through the air. Sometimes a spark is not seen until at a distance. It is owing to this inflammation that sparks kindle gunpowder. Were they nothing more than red-hot, they would cool in their passage, and fail in setting it on fire.

All iron contains charcoal, and a certain portion of oxygen, which gives it hardness, and without which, iron would be a soft metal. To the different proportions of these two substances, are owing the different qualities of iron.

Iron, at a high temperature, readily imbibes charcoal. The compound of charcoal and iron is the substance which flies off in brilliant sparks from iron of a white or high red heat, when struck by the hammer on the anvil. Different pieces of iron, heated to a red heat, unite perfectly by hammering, or are *welded*. Some silicious sand into the fire—the sand melts, by uniting with the oxydated metal, and forms a liquid glazing, which prevents further oxydation.

The medicinal preparations of iron (principally the sulphate) when properly administered, prove friendly to life. They excite a brisker circulation; inspire vigour and alacrity; impart a more healthy colour to the wan countenance; diffuse a general warmth over the whole system; and exercise an astonishing power in recruiting exhausted nature.

LEAD,

CALLED the metal of Saturn, was worked in very remote ages, in Attica, in England, and in Spain. Its colour is bluish white, very bright at first, but tarnishes on exposure to the air. Lead has never been found pure; its, ores are numerous, and occur in beds, and in veins, in all parts of the world; and, next to iron, is the most common of metalliferous ores. The most common lead ore is the sulphuret, called *Galena*, from a Greek word, signifying to shine, and from it are derived the immense quantities of lead used by man; galena yields lead, sulphur, oxide of iron, and sometimes lime and silex, and generally silver. The ore is first picked free from stony matter, after which, being broken up, it is roasted in large furnaces, by which the sulphur and arseuic are driven off; from thence it is carried to the smelting furnace, where it is thrown in, mixed with the coals, and the fire kept up at a strong red heat by the working of large double bellows. The ore now melts, and running through the coals, is reduced, the metallic lead being afterwards let out at the bottom of the furnace, and cast into pigs. Silver is separated from lead, by exposing the metal to a high heat, called cupellation,—the ore is laid on a cupel, or cup of bone-ash, and a current of air, by means of a blast furnace, passed over it; by which process the lead is oxidated, and converted into litharge, and the silver remains pure in the cupel. The litharge is afterwards reduced into reguline lead, by melting it with coals in a close furnace. Litharge, fused with salt, decomposes it; the oxide unites with the muriatic acid, and forms a yellow paint. Lead in combination with half its weight of tin, forms solder; with a quarter of its weight of antimony, it forms the metal of which printing-type is made. Oxide of lead becomes thick in oil, and is the basis of plasters, paints, &c.

The annual produce of the lead mines of Great Britain is supposed to be 46,000 tons. If lead ore yields 75 lbs. of metal out of 100, it is considered rich; if only 40, it is not worth working.

ZINC

WAS discovered in the 16th century, and is never found pure. Its ores are blende (called, by miners, black-jack), red zinc, electric calamine,

calamine, and white vitriol.   The ores of zinc are found in most mineral countries, accompanied by iron pyrites, sulphuret of lead, some of the ores of silver, and by calcareous spar and quartz.   Calamine is the most common of the zinc ores, and is found in lime-stone, in Derbyshire and Cornwall.   White vitriol is an hydrated sulphate of zinc and is generally found crystallized, and of a white colour ; it is a scarce variety, and is formed by the absorption of oxygen from the atmosphere, by the sulphuret.   Zinc has a considerable degree of lustre, but tarnishes very quickly, and is very ductile ; it is principally used for roofing, and for pipes ; but the readiness by which it is attacked by acids, prevents its being much used for culinary purposes.   It is also used in galvanism, and, combined with tin, forms an amalgam used for electrical machines.

The method of reducing the ores of zinc was for a long time unknown in this country, and it is stated that the process was obtained from the Chinese, by a person who went over for the purpose.   It is as follows : —The ore, having been brought out of the mine, is broken into small pieces, and submitted to a gentle heat in a reverberatory furnace, until the carbonic acid, sulphur, and other very volatile substances, are expelled.   After this, it is transferred to large jars made of baked clay, each of which has a pipe passing through the bottom, and rising nearly to the top of the vessel in the inside ; they are also furnished with covers, which are now luted on air-tight.   When six or eight of these jars are thus filled, they are placed in a long furnace, the lower end of each pipe passing through the iron floor, or grating, and dipping at their extremities into a vessel of water.   The fire is now lighted, and the zinc, after rising in sublimation, is obliged to descend through the pipe into the water, where it collects in a granular form.   The whole process is, therefore, merely that of distillation.

Although this metal is extremely brittle, it may easily be rolled into thin sheets, if previously heated to about 300 degrees of Fahrenheit. When melted, and run into hot water, and thus granulated, it is called spelter, and is used for hard soldering.

MERCURY,

CALLED after the heathen deity whose name it bears, was known to the ancients.   It is an exceedingly heavy metal, evaporating easily on exposure to a high temperature, on which account it is much employed in amalgamations.   It is always in a fluid state, and when submitted to a sufficient degree of cold, is similar in appearance to other metals, and may be beaten out into plates ; it is also so volatile, that it may be distilled like water ; and so elastic, when in a state of vapour, that it is capable of bursting the strongest vessels.   The congealation of mercury takes place at 39 degrees below zero.

G

The principal mines which furnish this metal are those in Idria, Deux Ports, Almaden, and Guanca Vellica : it is not peculiar to any one soil, being found in quartz, indurated clay, calcareous spar, &c. The ore, quicksilver, in the mines of Idria is dug out with picks ; drops of the pure liquid are to be seen all over the place, and inhaling the quicksilver vapour in the mines is extremely unwholesome, and the heat oppressing almost to suffocation. The smelting kilns of Idria yield from 600,000 to 700,000 pounds of mercury annually, of which 100,000 pounds are converted into cinnabar, or vermilion; into mild or corrosive sublimate ; and into precipitate.

The mode of reduction is as follows :—The ore, after it has been sorted, and those pieces which appear destitute of metal, thrown away, is ground to a coarse powder, and well mixed with from 1-8th to 1-4th its weight of fresh-slacked lime. Iron retorts, capable of holding 80 or 90 pounds of this mixture are charged, and placed in a long furnace, and heat gradually applied, until the whole of the metal has distilled over into receivers, connected with the beaks of the retorts, on the out-side. When the vessels are cool they are emptied of their contents, and again set to work as before. This is the process usually adopted with the cinnabars ; the lime, in this case, having a much stronger affinity for the sulphur than the mercury, combines with it, and fluid mercury results.

The mines of Almaden have been worked for 22 centuries, as Theophrastus speaks of their renowned cinnabar: their present ascer-tained contents are still so great, that they will afford a yearly supply of 22,000 quintals for at least 500 years to come.

COBALT

Is not found in its native state ; but, in Cornwall, is combined with arsenic, sulphur, and iron ; occurring in veins accompanying ores of copper, native bismuth, and silver.

The greatest part of the cobalt seen in this country comes from the Saxon mines, under the form of zaffre. Cobalt is much used in the arts, particularly in the manufacture of porcelain, as many beautiful blues are extracted from it. All blue glass is coloured with cobalt. The colour of the ores are, silver and tin white, steel grey, straw yellow, flesh red, crimson, brown, and black ; and are mostly soft or brittle. They are found abundantly in Cornwall, at St. Columb, in Illogan, in St. Just, and Gwinear.

MANGANESE

Is never found pure, but combined with oxide of iron, sulphur, the sulphuric or carbonic acid, and barytes, but mostly with oxygen. From the black oxide of manganese, the oxygen gas, used by chemists, is obtained ; as well as almost all the oxygen consumed in bleaching, in Great Britain, France, and Germany.   The ores are found abundantly in Cornwall, Ireland, &c. ; first found in Devon, about 1770, at Upton Pyne, near Exeter, the ores from which mine, with two smaller ones, upon the same lode, for many years supplied the United Kingdom with this article.   The oxide of manganese exists in the vegetable kingdom, being found generally in the ashes of plants.   It is used at every potter's kiln, to give the dark glazing to the coarsest earthenware.

Analysis.—Digest 100 grains of the ore with the application of heat, a sufficient quantity of moderately strong muriatic acid, until nothing remains undissolved but a white powder—which being washed, dried, &c., is the silex of the ore.   Add caustic ammonia to the clear solution, so long as any precipitate is given, which will precipitate the iron contained in the ore ; separate this by a filter, wash, dry, and reduce to the magnetic state.   Put all the liquors together, evaporate to dryness, and wash the residue with distilled water—the powder now remaining, is the oxide of manganese contained in the ore.

ANTIMONY.

THE date of the discovery of antimony is unknown, but it was previous to the 16th century.   Metallic antimony, the regulus of commerce, is of a bluish white colour ; its principal use, in the arts, is to soften metals, as lead, tin, &c.

Native or pure antimony is found in veins, in the mountains of Dauphine, in the Hartz, and in Sweden, disseminated in calcareous spar.   The ores are five in number, combined with oxide of iron, cobalt, arsenic, silex, sulphur, and oxygen, and found principally in veins, in primitive or the older secondary mountains of Sweden, Saxony, France, Bohemia, and England.

The antimony of commerce is principally extracted from the sulphuret.

Its medicinal properties, which are valuable, are stated to have been discovered by accident.   A German monk, having mixed some repeatedly with the food of swine, thought they grew fat upon it.   He thereupon, determined to try its effects upon some of his brethren, with whom it succeeded fatally !

Large quantities of antimony are raised in Cornwall and Devon, principally in the neighbourhood of Endellyon, St. Stephens, and St. Austle.

PLATINA.

PLATINA was discovered in 1735, by Antonio de Ulloa, and was known previous to that time, under the name of White Silver.

It is the heaviest of metals, and is very malleable and flexible.   It is

employed in the construction of mirrors and reflectors, and is in much
request for mathematical instruments which require precision.

Platina is only found in grains,—and in the sands which contain it,
are found iron, copper, chrome, titanium, iridium, osmium, rhodium,
and palladium. It is never found pure, is with difficulty fused, and
can only be refined with great care ; consequently, it bears a high
price, often higher than gold. It is found chiefly in Russia and
America.

NICKEL

Is found very abundantly in Germany, where it occurs in the form of a
sulphuret, of a pale copper-red colour, called Kupfernickel. The ores
of nickel are few, and generally accompanied by ores of silver and
cobalt, by calcareous spar and quartz, and other substances. It is one
of the least abundant metals, and found in France, Spain, Saxony,
Bohemia, and in Cornwall, especially at Pengelly Mine, in St. Ewe,
and at Wheal Chance, near St. Austle. The thorough purification of
this metal is very difficult ; it is magnetical, and acquires polarity by
the touch ; it is fusible under an intense heat, after which it is malleable, and assumes a white colour, between that of silver and tin. Its
alloy with copper possesses a superiority, in point of durability of
colour, over that of plated metal, and is the chief ingredient of German
silver.

ARSENIC.

Is of a bluish white ; it is found nearly pure, being alloyed only by
small portions of iron, and sometimes of gold and silver, in primitive
mountains, in veins accompanied by ores of silver, cobalt, lead ; by
calcareous spar, fluate of lime, and quartz.

It is principally found in Germany, Sicily, England, and Ireland.

Arsenic, combined with Iron, forms arsenical pyrites, or mis-pickel,
in some varieties of which, silver is found. It is also found combined
with 25 parts of sulphur, forming an ore of a red or orange colour,
called realgar, and, with 43 parts sulphur, it forms a bright yellow-
coloured ore, called orpiment.

The pyrites of arsenic are very common in the Cornish mines ; and
the oxide is obtained in great abundance from the desiccation of the
ores of tin. The white oxide of arsenic, which is the arsenic of commerce, is often obtained from it. The beautifully-twisted scrolls in
the stalks of old wine-glasses, are arsenic.

A GLOSSARY OF THE TERMS USED IN MINING, &c.

Adventurers.—Those who have shares in a mine, in contradistinction to the lord who is owner of the soil.

Adit Level.—A horizontal excavation, through which the water drawn thereto by the engine, and that which falls from above, passes off to the surface; this level is usually commenced from the bottom of the deepest neighbouring vale, and extended through a great part of the mine; the top adit is the adit first driven; the deep adit, the lowest level driven. *An adit* is the adit driven purposely for the ventilating, watering, or draining the mine.

Air Machine.—An apparatus for forcing fresh air or withdrawing foul air from places badly ventilated.

Air Pipes.—Tubes, or pipes of iron or wood, for ventilating under ground, or for the conveyance of fresh air into levels having but one communication with the atmosphere, and consequently, no current of air.

Arched.—The roads in a mine when built with stones or bricks are generally arched level drifts.

Attle.—Rubbish, deads, refuse, or stony matter.

Arch.—A piece of ground left unworked.

Average Produce.—The quantity of fine copper contained in 100 parts of ore; thus, a parcel of ore having a produce of 10⅝ contains 10⅝ per cent. of fine copper, being rather above the general average of the copper ores of Cornwall.

Average Standard.—The price per ton of fine copper in the ore, after deducting returning charges for smelting, of £2. 15s. per ton of ore in Cornwall, and £2. 5s. per ton of ore at Swansea. The regulation of the standard depends entirely on the price which fine copper bears in the market, rising and falling in the same proportion. Supposing the produce of a parcel of ore to be 10, and the price at which it was sold to the smelter to be £8. 18s., the standard of that parcel will be thus obtained :—ten tons of the ore will be required to yield one ton of fine copper; therefore, £8 18s. ✕ 10 = £89. will be the value of the ore containing a ton of metal. For the same reason the returning charge of £2. 15s. must be multiplied by 10, making £27. 10s., which, added to the former sum of £89., makes £116. 10s. —being the standard of that parcel. Low produce ore will naturally have a higher standard.

Bal.—The miners' term for a mine.

Bob.—The engine beam.

Back.—The back of a lode is the part of it nearest the surface; the back of a level is that part of the lode extending above it to within a short distance of the level above.

Blasting.—A hole is made with a borer, into which gunpowder is inserted, and being confined and set fire to, it forces off a portion of the rock or lode.

Bed.—A seam, or horizontal vein of ore.

Bunch, or Squat.—A quantity of ore of small extent, more than a stone, and not so much as a course; a mine is said to be *bunchy* when these are found in place of a regular lode.

Branch.—A small vein which separates from the lode and frequently again unites with it; the term is also applied to a string of ore falling into the lode.

Batch.—A certain quantity of ore sent to the surface by any pair of men.

Blast.—The air introduced into a furnace.

Bounds.—The proprietary of tin ore over a given district.

Breast.—The face of coal workings.

Buddle.—A frame, made of wood and filled with water, to wash lead ore.

Burrow.—A heap of deads, attle, rubbish, &c.

Burning House.—The furnace in which tin ores are calcined, to sublime the sulphur from pyrites; the latter, being thus decomposed, are more readily removed by washing.

Bar of Ground.—Any course of vein which runs across a lode, or different from those in its vicinity.

Beat Away.—To excavate, usually applied to hard ground.

Bend.—Indurated clay. a name given by miners to any indurated argillaceous substance.

Borer.—A boring instrument, with a piece of steel at the end, called a boring belt

Bucking Iron.—The tool with which the ore is pulverized.

Buckers.—Bruisers of the ore.

Bucking Plate.—An iron plate on which the ore is placed for being sucked.

Black Tin.—Tin ore, triturated and washed for smelting.

Blower.—A smelter.

Burden.—The tops or deads of " stream work" that lie over the stream of tin, and which must be first cleaned.

Blende.—One of the ores of zinc, composed of iron, zinc, sulphur, silex and water; on being scratched, it emits a phosphoric light.

Black Jack.—Blende.

Country.—The strata through which the lodes pass.

Counter Lode.—A lode which includes a considerable angle with the direction of the other lodes in its vicinity, or runs different from them.

Capel.—A stone composed of quartz, schorl, and hornblende, usually occurring in one or both walls of a lode, and more frequently accompanying tin than copper ores.

Cross Cut.—A level driven at right angles with the direction of the lode.

Course of Ore.—A portion of the lode, containing a regular vein.

Crushing.—Grinding ores without water.

Cutting.—An air course set up at either end of the work after the ore has been brought out.

Costeaning.—The discovering lodes, by sinking pits in their vicinity and drawing transversely to their supposed direction.

Cob.—To break the ores with hammers in such a way as to separate the dead or worthless parts.

Cross Course.—A lode or vein which intersects or crosses a lode at various angles, and generally throws the lode out of its proper course.

Cathead—A small capstan.

Cope.—To agree to get lead ore at a fixed sum per dish, or load, or other measure.

Clack.—The valve of a pump of any description.

Casing.—A division of wood planks separating a footway, or a whim or engine shaft from one another.

Collar of a Shaft.—The timber by which its upper parts are prevented from falling.

Crop.—The best ore.

Cank.—Whimstone.

Charger.—An instrument in form of a carpenter's augur, for charging holes for blasting, which are dug horizontally.

Cofer.—Cofering is beating a quantity of clay round the backing in shaft, to prevent the water coming through, and to hold it back in strata.

Cage of a Whim.—The barrel on which the rope is wound up.

Crib, or Curb.—A circular frame of wood, screwed together, as a foundation for bucking or pulverising ore in a shaft.

Coffin.—Workings open, like an intrenchment, having no shaft.

Carbona.—A dropper from a lode in irregular masses.

Chimming.—Dressing tin ores—scarcely differing from tossing or tozing.

Dradge.—Ore mixed with other substances.

Dissueing.—If the lode is small and rich, they commonly only break down the country on one side of it, by which the lode is laid bare and may be taken down free from waste.

Driving.—Cutting and blasting horizontally.

Dropper of the Lode.—A branch when it leaves the main lode.

Dead Ground.—A portion of the lode in which there is no ore.

Drift.—An excavation made for a road underground.

Durns.—A frame of timber with boards placed behind it to keep open the ground in shafts, levels, &c.

Deads.—Attle, or rubbish.

Draft Engine.—An engine used for pumping.

Dish, or Dues.—That portion of the produce of the mine paid to the landowner or lord.

Dressers.—Cleaners of Ore.

Dowsing Rod.—The hazel rod of divination, by which some pretend to discover lodes.

Elvan.—A course of porphyry and claystone.

End.—An adit is said to be driven end,when it is in a line with the grain of a coal.

Engine Shaft.—The pit or shaft, by which the water is drawn by the engine from the lower parts of the mine to the adit or surface.

Elve.—The shaft or handle of a pick.

Fathom.—Six feet.

Fluccan.—A soft clayey substance, generally found to accompany the cross courses and slides, and occasionally the lodes themselves, but *fluccan,* when applied to a vein, means a cross one, or course composed of clay.

Forks.—Pieces of wood used to keep up the sides in some places ; a mine is said to be *in fork,* when the engine keeps out the water

Flat Rods.—For communicating motion from the engine horizontally to other shafts at some distance, which are called *flat rod shafts.*

Fault.—An intersection of the strata.

Feeder.—A branch when it falls into the lodes.

Footway.—Is a wall under the lode ; also the ladders by which the workmen ascend and descend.

Fluke.—The head of the charger ; an instrument used for cleansing the hole previous to blasting.

Fang. - A niche cut in the side of an adit, or shaft, to serve as an air course ; sometimes a main of wood pipes is denominated a fanging.

Floor.—A bed of ore in a lode, supposed not to continue to any great depth ; a strata of ore.

Gozzan.—Oxide of iron and quartz, generally occurring at shallow depths, and considered by miners a favourable indication of ore at deeper levels.

Grain Tin.—The finest tin, smelted with charcoal.

Gad.—A wedge.

Gal.—Rusty iron ore

Gatches. - The after leavings of tin.

Grass.—The surface of the mine ; ores are said to be *at grass* when upon the surface.

Ground.—The country; the stratum in which the lode is found.

Grinder.—Machinery for crushing.

Gowan.—Decomposed granite, but sometimes applied to the solid rock.

Gudgeon.—A pin working in a nozzle in a steam engine.

Gate.—Road or way underground ; it has various uses, either for water, air, or for bringing out of the mine, coal, &c. &c.

Huel.—The ancient name for "mine," corrupted into " wheal."

Halvans or Hennaways.—Ores not sufficiently rich to be offered for sale, and like the *attle* are generally left on the surface.

Heave.—The horizontal dislocation which occurs when one lode is intersected by another having a different direction.

Hade.—The underlay or inclination of the vein.

Hanging Wall.—The wall or side over the lode.

Horse.—The dead ground included between two branches of a lode at the joint of their separation.

Hutch.—Cistern, or box.

Hauling.—Drawing ore or attle out of the mine.

Hulk.—An old excavated working ; to hulk a lode is to pick that out so far as they can reach, and blow down the rest with gunpowder.

Ironstone.—Hard, bluish grey stone, keeping its course like a lode.

Jumper.—A long borer used by one person.

Jigger.—A cleaner of ores.

Jigging.—Sifting copper and lead ores under water.

Killas.—A clay slate, occurring in different parts of the mine.

Kibble.—A bucket usually made of iron, in which the ores, &c. are usually drawn to the surface

Knits.—Small particles of lead ore.

Kal.—Hard.

Karer.—A sieve.

Knockings.—Lead ore, with spar, as cut from the veins.

Keeve.—A large vat.

Kibble Filler.—Man who sends up work to the surface.

Knocked.—Abandoned.

Lode.—A regular vein producing ore affording any kind of metal.

Leader.—A branch, or small vein of the main lode.

Levels—are driven on the lode, usually 10, 20, 30, 40, &c., fathoms below the adit.

Ley.—Standard of metal; contents in pure metal.

Loobs.—Slime containing ore.

Lama.—Slime or schelm for the amalgamation.

Leavings.—The ores which are left, after the crop is taken out.

Lost Levels.—Levels which are not driven horizontally.

Lode Plot.—A lode that underlies over fast, or horizontal, and may rather be called a flat lode.

Launders.—Tubes or gutters for the conveyance of water, their form being that of a long box, wanting the upper side of both ends.

Leat.—A water course.

Lot.—A certain proportion taken as dues for the Lord of the Manor.

Lander.—Man who attends at the mouth of a shaft to receive the kibble, in which ores, rubbish, &c. are brought to the surface.

Lofty tin.—Massive and superior tin.

Mundic.—Iron pyrites, arsenic and sulphur, found very abundantly in Cornish and Irish mines, and often laid aside as useless, when in fact it might be turned to good account for sulphur.

Mad-water.—Water drawn from the pump, and which returns again from whence it was drawn, through bad management.

Mineral.—Ore, &c.

Mear.—Thirty-two yards of ground in the vein.

Monton.—A heap of ore, a batch under process of amalgamation, varying in different mining districts.

Moor.—A root, or a quantity of ore in a particular part of the lode.

Meat-earth.—The cultivated soil.

Needle or Nail.—.—A long taper piece of copper or iron, with a copper point, used when stamping the hole for blasting, to make by its insertion an aperture for a fusee or train.

Nogs or Noys.—Square pieces of wood, piled on each other to support the roof of a mine.

Old Man.—Places worked centuries ago.

Owners Account Men.—Workmen paid at so much a day.

Open Cast.—When a vein is worked open to the day, or surface.

Produce.—See " average produce."

Pillion.—The tin that remains in the slags after it is first melted.

Peach-stone.—A blueish green soft stone.

Pack.—A quantity of material, either wood or coals, is placed or piled up to support the roof, or for other purposes.

Plat.—Ground taken away to contain any ore or deads.

Pricker.—A thin piece of iron used to make a hole for the fusee, or match to fire a blast.

Point of Horse.—The spot where the vein is divided into one or more branches.

Parcel.—A heap of ore dressed and ready for sale.

Pillar.—A piece of ground left to support the roof or hanging wall.

Pitwork.—The pumps and other apparatus of the engine shaft.

Poldavy.—Coarse sacking for coal sacks, &c. &c.

Pol-roz, pronounced *polrose.*—The pit underneath a waterwheel.

Prill.—A solid piece of ore; a specimen.

Punch.—A piece of timber used as a support for the roof.

Pitman.—One employed to look after the lift of pumps and the drainage.

Purser.—The cashier, or paymaster, at the mines.

Para.—A gang or party of men.

Pick.—An instrument in common use, as well in agriculture as in mining.

Pipe.—A vein running unlike a rake, having a rock root and sole.

Pryan—That which is productive of ore, but does not break in large stones, but only in pebbles, with a mixture of clay.

Quears.—Crevices in lodes.

Rising.—Digging upwards.

Rib.—A pillar of coal left as support for roof.

Racking.—A process of separating small ores from the earthy particles by means of an inclined wood frame, the impurities being washed off, and the ore remaining near the head of the rock taken from thence undergoes tossing.

Rubble.—Loose stones.

Roof.—The part above the miner's head, that part of the strata lying immediately upon the coal.

Run.—When excavations fall together.

Rack.—An inclined plane on which the ores and slime are washed and separated.

Ratchell.—Loose stones.

Rake.—An oblique vein.

Run of a Lode.—Its direction.

Standard.—See " average standard."

Stem.—A day's work.

Shift.—The time a miner works in one day.

Shaft.—A sinking or pit, either on the lode or through the country.

Slide.—A vein of clay which, intersecting a lode, causes a dislocation vertically.

Stamps—Machinery for crushing ores.

Stopping.—Cutting down mineral ground with a pick.

Stull.—Timber placed in the backs of levels, and covered with boards or small piles to support rubbish.

Sump.—A pit sunk in the engine shaft below the lowest workings.

Seam.—A horse load.

Sett.—A mine, or number of mines, taken upon lease.

Sough.—An adit or level for carrying off the water.

Slickings.—Narrow veins of ore.

String.—A small vein.

Scraper.—A piece of iron used to take out the pulverized matter which remains in the hole when bored previous to blasting.

Slimes.—Mud containing metallic ores, mud or earthy particles mixed with the ores.

Stampheads.—The iron weight or head connected with the stamps.

Scovanlode.—A lode having no gozzan on its back, or near the surface.

Shieve.—The pulling over which the whim rope passes.

Sinking.—Digging downwards ;—in rising and sinking a shaft, one set of men sink from a certain level, while another set rise from a lower level to meet them.

Spalling.—Breaking up into small pieces for the sake of easily separating the ore from the rock, after which it undergoes the process of cobbing.

Spend.—To break ground, to work away.

Sturt.—When a tributer takes a pitch at a high tribute, and cuts a course of ore, he sometimes gets two, three, to five hundred pounds in two months, this great wages is called a sturt.

Streamers.—The persons who work in search of stream tin.

Shears.—Two very high pieces of wood placed in nearly a vertical position in each side of a shaft, and united at the top, over which, by means of a pulley, passes the capstan rope ; this is for the convenience of lifting out or lowering into the shaft, timber, or other things of great length and weight.

Sump Shaft.—The engine shaft.

Scrin.—A small vein.

Sole,—The seat or bottom of the mine, applied to horizontal veins or lodes.

Stempls.—Wood placed to go up and down the mine instead of steps.

Tribute.—A proportion of the ore which the workman has for his labour. Tributers generally work in gangs, and have a limited portion of a lode set them, called a "tribute pitch," beyond which they are not permitted to work, and for which they receive a certain portion of the ore, or so much in the pound as agreed upon, in value of what they raise, the workmen finding all tools, candles, and gunpowder, used in working their pitch. The men are thus interested in preserving everything of any value found in prosecuting the works, and all collusions are avoided about rates of wages. The general custom is, to set the work to be done to certain contractors, who, by themselves or labourers employed by them, extract the ores ; the extent of a pitch usually being 10, 15, or 20 fathoms of level, and 10 fathoms in depth.

Tutwork.—Where the labourer earns in proportion to his labour, being paid for drawing or sinking at a certain price per fathom.

Tackle.—Windlass rope and kibble.

Ticketings.—The weekly sales of ores.   The adventurers, or their agents meet together at 12 o'clock, and, whilst sitting round a table, each buyer gives in his ticket, offering a certain sum per ton for so many tons of ore.   The tickets are then read aloud by the chairman (who is always the agent, having the largest quantity of ore for sale for the day) and the persons present, note the prices offered, the lots or different samplings being sold to the highest bidder.   After the sales, the parties dine together at the expense of their respective mines.

Ton.—A ton in weight varies in different districts.   The common ton is 20 cwts. of 112 lbs. or 2, 240 lbs.   In Cornwall, the miners ton is 21 cwt. of 112 lbs. or 2,352 lbs.

Troubles.—Faults or interruptions in the stratum.

Tossing, Tozing, or Terloobing.—A process consisting in suspending the ores by violent agitation in water, their subsidence being accelerated by packing ; the lighter and worthless matter remains uppermost.

Tying.—Washing.

Tugs.—Hoops of iron fastened to the covers to which the tackles are affixed.

Trade.—Attle or rubbish.

Thurl.—A long adit in a coal pit.

Tamping.—The material, usually soft stone, placed upon the gunpowder in order to confine its force, which would otherwise pass up the hole ; also the process of placing the material.

Tummals.—A great quantity ; a heap.

Timberman.—The man employed in placing supports of timber in the mine.

Tamping iron.—Tool used for beating down the earthy substance in the charge used for blasting.

Trunking.—Process of extracting ores from the slimes, the ores subsequently undergo the process of racking and tossing.

Tub.—A cast iron cylinder put in the shaft instead of bricking, for the purpose of beating out the water, and making it rise to a level.

Underlay shaft.—Shaft sunk on the course of a lode.

Underlayer.—A perpendicular shaft sunk to cut the lode, at any required depth.

Underlay.—When a vein hides, or inclines from a perpendicular line, it is said to underlay.

Vanning.—Removing the impurities from tin ore.

Vein.—Any substance different from the rock ;   A rake vein is perpendicular or nearly so, a pipe vein, nearly horizontal.

Van.—To wash or cleanse a small portion of ore on a shoal.

Vugh, or Vogle.—A cavity.

Vat.—A wooden tub, used to wash ores and mineral substances in.

Wheal.—The ancient Cornish called a mine " huel," which has been corrupted by the moderns into "Wheal."

Winch, or Winse.—The wheel and axle frequently used to draw water, &c. in a kibble by a rope.

Winding Engine.—One used to draw up ore, attle, &c.

Weighboard.—Clay intersecting the vein.

Working Big.—Sufficiently large for a man to work in.

Whim Shaft.—The shaft by which the stuff is drawn out of the mine by horse or steam whim.

Whim.—A machine used for raising ores, &c. worked by horse, steam, water, &c.

Work.—Ores before they are cleaned and dressed.

Wastes.—Vacant places left in the gobbing, in each side of which the rubbish is packed up for the better support of the roof.

Winze.—A sinking on the lode communicating with one level, for proving the lode, or for ventilating the drivings.

Wipers.—The cogs of a horizontal axle, moved either by water or steam.

Zyghyr, or Sigger.—When a slow stream of water issues through a cranny, it is said to sigger, or zyghyr.

LIST OF SUBSCRIBERS.

<table>
<tr><td></td><td align="right">COPIES</td></tr>
<tr><td>Ashe, Lieut.-Colonel Wellesley, of Ashencourt . . . .</td><td align="right">4</td></tr>
<tr><td>Abbott, William, Esq., Wyndham Place . . . .</td><td align="right">1</td></tr>
<tr><td>Abbott, William, Esq., Jun., Do. . . . . . .</td><td align="right">1</td></tr>
<tr><td>Aimè, Benjamin, Esq, Belmont Terrace . . . . . .</td><td align="right">4</td></tr>
<tr><td>Brookhouse, Martin, Esq., Reigate . . . .</td><td align="right">3</td></tr>
<tr><td>Brown, Joseph, Esq., Leicester . . .</td><td align="right">1</td></tr>
<tr><td>Blamey, Francis, Esq., Gwennap, Cornwall . .</td><td align="right">1</td></tr>
<tr><td>Burgess, William, Esq., Trengrove, Cornwall</td><td align="right">1</td></tr>
<tr><td>Buxton, Samuel, Esq., St. Mildred's Court . . .</td><td align="right">1</td></tr>
<tr><td>Bankart, S. S., Esq., Leicester . .</td><td align="right">1</td></tr>
<tr><td>Banner, Edward Gregson, Esq., County Terrace . . .</td><td align="right">1</td></tr>
<tr><td>Birdsey, Mr. William, London</td><td align="right">5</td></tr>
<tr><td>Blake, Mr., Liskeard, Cornwall</td><td align="right">1</td></tr>
<tr><td>Clarke, The Rev. Charles, Queen's College, Cambridge . .</td><td align="right">1</td></tr>
<tr><td>Chippindale, William, Esq., Bunhill Row . . . .</td><td align="right">1.</td></tr>
<tr><td>Chippindale, Charles, Esq., St. John's Wood . . . .</td><td align="right">1</td></tr>
<tr><td>Curry, Thomas, Esq., Bond Court . .</td><td align="right">1</td></tr>
<tr><td>Cox, Thomas Baker, Esq., Poultry</td><td align="right">2</td></tr>
<tr><td>Cobre Mining Association, Austin Friars . , . .</td><td align="right">1</td></tr>
<tr><td>Cardozo, Samuel, Esq., Redruth, Cornwall . . . .</td><td align="right">1</td></tr>
<tr><td>Coward, Noah, Esq., Gunnis Lake, Tavistock . . . .</td><td align="right">1</td></tr>
<tr><td>Clamp, Robert Bircham, Esq., Ipswich . . .</td><td align="right">1</td></tr>
<tr><td>Cuell, William Henry, Esq., Clapham . .</td><td align="right">1</td></tr>
<tr><td>Cooper, Mr. John Moore, Brixton . . .</td><td align="right">1</td></tr>
<tr><td>Cooper, Mr. Thomas White, Princes Street . .</td><td align="right">1</td></tr>
<tr><td>Dickin, George, Esq., Moreton Hall, Salop . .</td><td align="right">1</td></tr>
<tr><td>Devonshire, John Kempe, Esq., St. Swithin's Lane .</td><td align="right">1</td></tr>
<tr><td>Edwards, John, Esq., Upper Stamford Street . ,</td><td align="right">1</td></tr>
<tr><td>Eives, John, Esq., Newington, Surrey .</td><td align="right">1</td></tr>
<tr><td>Flight, Thomas, Esq., Bond Court-House . . .</td><td align="right">1</td></tr>
<tr><td>Firmin, The Rev. John Palmer, B.A., Davenham. Cheshire . .</td><td align="right">1</td></tr>
<tr><td>Firman, The Rev. Frederick, B.A., Boston, Lincolnshire . .</td><td align="right">1</td></tr>
<tr><td>Freeman, Robert, Esq., Swanton Morley, Norfolk . .</td><td align="right">3</td></tr>
<tr><td>Freeman, John, Esq., Rudham, Norfolk . . . .</td><td align="right">1</td></tr>
<tr><td>Freeman, Joshua, Esq., Dersingham, Norfolk . .</td><td align="right">1</td></tr>
<tr><td>Goodall, Charles, Esq., City Club-House . . . .</td><td align="right">2</td></tr>
<tr><td>Herapath, John, Esq., Proprietor of Railway Magazine and Commercial Journal</td><td align="right">3</td></tr>
<tr><td>Hacket, Thomas, Esq., Birchin Lane . . . . .</td><td align="right">2</td></tr>
<tr><td>Herron, James, Esq., London . . . . .</td><td align="right">1</td></tr>
<tr><td>Hudson, Miss, Regent's Park . . . .</td><td align="right">1</td></tr>
<tr><td>Hudson, Mr. Thomas, Jun., Carlisle . . . .</td><td align="right">1</td></tr>
<tr><td>Hill, Thomas Irving, Esq., London . . . .</td><td align="right">1</td></tr>
<tr><td>Hill, Thomas, Esq., Ingram Court . . .</td><td align="right">1</td></tr>
<tr><td>Hitchcock, Samuel, Esq., Solicitor, Mistley, Essex . . . .</td><td align="right">1</td></tr>
<tr><td>Heseltine, Edward, Esq., Birchin Lane . . .</td><td align="right">1</td></tr>
<tr><td>Howie, John, Esq., Park Place, Highbury . .</td><td align="right">1</td></tr>
<tr><td>Harrison, G. W., Esq., 75, Old Broad Street . .</td><td align="right">1</td></tr>
<tr><td>Hollyer, Mr. M. H., York Place, Kentish Town .</td><td align="right">1</td></tr>
<tr><td>Hitchens, Mr. Josiah H., Tavistock, Devon .</td><td align="right">1</td></tr>
<tr><td>Hennah, Mr. William, South Caradon Mine, Cornwall</td><td align="right">1</td></tr>
</table>

COPIES.

Innes, Mr. James, Cheapside 1

Jago, Mr., Liskeard, Cornwall 1

Kitto, Captain Thomas, Junior, Redruth, Cornwall 1
Knotwell, Captain, Redruth, Cornwall 1

Large, Robert, Esq., Great Clacton, Essex 1
Linton, William, Esq., St. Osyth, Essex 1
Large, Mr. Henry, Great Clacton 1

Mount, William, Esq., London 4
Murray, Adam, Esq., Parliament Street 1
Markwell, William, J., Esq., North and South American Coffee House . . 1
Matthews, Oliver H., Esq., St. Day, Cornwall 4
Matthews, Joseph, Esq., Mining Office, Tavistock 1
Malachy, Joseph, Esq., Calstock, Cornwall 1

Powell, John Richard, Esq., Preesgwene House, Chirk 1
Puckey, Mr. John, Fowey Consols Mine, Cornwall 1
Pike, Robert Hart, Esq., Rectory House, London Wall . . . 1

Richards, William, Esq., Redruth, Cornwall 1
Rule, Alfred, Esq., London 1
Reynolds, Captain William, Redruth, Cornwall 2

Sweny, Mark Halpen, Esq., Capt. R.N., Naval Club-House 1
Stainsby, Peter, Esq., Finsbury Square 1
Sims, Mr. William, Junior, Redruth Cornwall 1

Turner, Edmund, Esq., M.P., Truro, Cornwall 1
Turner, Charles Walsingham, Esq., H.M.'s Consul at Carthagena, Spain . 1
Territt, Mrs. Regent's Park . . . , . . . 4
Trenery, William, Jun., Esq., 50, Threadneedle Street . . . 4
Thomas, Mr. Stephen, London 1

Vivian, Captain John, Gwennap, Cornwall 1

Watson, William Henry, Esq., M.P., Temple . . 4
Watson, William, Esq., St. Osyth, Colchester , , . . 4
Watson, Lieut. John, Ballymacomick . , . . 1
Watson, Mr. William Howard, Lombard Street 1
Wheatley, Joseph, Esq., Leicester 1
Warton, Charles, Esq., 38, Threadneedle Street 1
Wye, Messrs., Brothers, Crutched Friars 1
Weir, Mr. William, Carlisle 1

MUNRO AND CONGREVE, Printers, 26, Duke Street, Lincoln's Inn Fields.